Testing and Modeling of Cellular Materials

Testing and Modeling of Cellular Materials discusses the characterization of cellular lattices through quasi-static and dynamic testing for use in light-weighting or energy-absorbing applications. Covering cellular materials, specifically additively manufactured lattices, this book further progresses into dynamic testing and modeling techniques for computational simulations. It presents modeling and simulation techniques used for cellular materials and evaluates them against experimental results to illustrate the material response under various conditions. The book also includes a case study of high-velocity impact that highlights the high strain rate effects on the cellular lattices.

Features:

- Covers different testing techniques used in quasi-static and dynamic material characterization of cellular materials.
- Discusses additive manufacturing techniques for lattice specimen fabrication.
- Analyzes different finite element modeling techniques for quasi-static and dynamic loading conditions.
- Presents a comparison and development of a phenomenological material model for use in computational analysis at various loading rates.
- Explores impact stress wave analysis under high-velocity loading.

The book will be useful for researchers and engineers working in the field of materials modeling and mechanics of materials.

Testing and Modeling of Cellular Materials

Derek G. Spear
Anthony N. Palazotto

CRC Press
Taylor & Francis Group
Boca Raton London New York

CRC Press is an imprint of the
Taylor & Francis Group, an **informa** business

First edition published 2023
by CRC Press
6000 Broken Sound Parkway NW, Suite 300, Boca Raton, FL 33487-2742

and by CRC Press
4 Park Square, Milton Park, Abingdon, Oxon, OX14 4RN

CRC Press is an imprint of Taylor & Francis Group, LLC

ISBN: 978-1-032-29022-5 (hbk)
ISBN: 978-1-032-29024-9 (pbk)
ISBN: 978-1-003-29963-9 (ebk)

DOI: 10.1201/9781003299639

Typeset in Times
by codeMantra

Contents

List of Figures

Preface

Additively manufactured metal lattices are a promising engineering material for incorporation into aerospace structures because of their high strength-to-weight ratio and energy absorption capabilities. This research aims to define the time-dependent characteristics of lattice structures and evaluate their ability to control a stress wave during high-velocity impact. Compression tests were conducted at various strain rates, from quasi-static to highly dynamic, and with a variety of lattice designs and configurations. These tests examined how the mechanical behavior of the lattice changed based on the lattice design properties and across the applied strain rates. The modulus of elasticity, yield strength, plateau stress, and toughness were observed to decrease at higher strain rates, revealing that the lattice designs exhibit a negative strain rate sensitivity. The unique nature of the mechanical response of the lattices required the development of a novel flow stress model that could be incorporated into a computer model for simulation. This new constitutive model was evaluated under three different test regimes, closely matching the results from the quasi-static, Split Hopkinson Pressure Bar, and Taylor impact tests. Experimental impact and simulation results indicated that the inclusion of a lattice core to a projectile would decrease the effects of impact wave propagation within the projectile. However, the decrease in projectile mass due to the inclusion of a less dense lattice section negatively influenced the projectile's penetration depth into a concrete target.

MATLAB® is a registered trademark of The MathWorks, Inc. For product information, please contact:

The MathWorks, Inc.
3 Apple Hill Drive
Natick, MA 01760-2098 USA
Tel: 508-647-7000
Fax: 508-647-7001
E-mail: info@mathworks.com
Web: www.mathworks.com

Authors

Derek G. Spear completed his MS in Aerospace Engineering at the North Carolina State University in Raleigh, NC, and his MS in Aviation Systems at the University of Tennessee in Knoxville, TN. He obtained his PhD in Aeronautical Engineering from the Air Force Institute of Technology, Wright-Patterson Air Force Base, OH. He has assumed various roles as executive officer/instructor pilot, flight commander/program manager, test pilot, helicopter flight commander, flight safety officer, director of operations, and commander. He has received many awards, honors, and decorations for his services. He has published white papers and presented his research work at international conferences.

Anthony N. Palazotto is a Distinguished Professor, Aerospace Engineering at the Air Force Institute of Technology. He has got over 50 years of experience in administration, research, and education within an Engineering College. He supervised and administered faculty within a departmental division. He developed research efforts and course offerings in composite materials, wave mechanics, elasticity, shells, fracture mechanics, and finite element techniques. He was highly involved in technical society activity as chairperson and founder of various committees. He was the author of over 700 presentations and publications, 271 are archival. He was the coauthor of a textbook entitled, "Nonlinear Analysis of Shell Structures," published by AIAA Educational Series, 1992, and invited chapters in "Additive Manufactured Handbook: Product Development for the Defense Industry," 2017, Registered Professional Engineer — Ohio (initial registration in New York). He delivered invited lectures at several academic institutions. Advised 22 Post Docs, 39 PhD's, and over 190 MS theses. He obtained over $5.75 million in funded research. He received the Outstanding Engineer and Scientist Award from the Affiliate Society of Dayton in 2011 and appointed as Distinguished Professor of Aerospace Engineering in 2011. He was a Fellow in AIAA, ASCE, Engineering Mechanics Institute, and American Academy of Mechanics. He was nominated for AIAA Crichlow Award, Honorary AIAA Fellow, and NAE. Granted patent number 10,843,783 Issue date Nov 24, 2020. He received, with his students, the best paper awards in 1982 Hetanyi Award, ASCE Aerospace Journal 2004, the AIAA Collier Research Award 2021 best paper.

Acronyms

AFIT	Air Force Institute of Technology.
AIAA	American Institute of Aeronautics and Astronautics.
AM	additive manufacturing.
ANOVA	Analysis of Variance.
ASTM	American Society for Testing and Materials.
BESO	Bi-directional Evolutionary Structural Optimization.
CAD	computer-aided design.
CDTI	Central Difference Time Integration.
CRD	Completely Randomized Design.
DEM	discrete element method.
DOE	Design of Experiments.
EBM	electron-beam melting.
EOS	equation of state.
ER	evolutionary volume ratio.
FEA	finite element analysis.
FEM	finite element method.
GE	General Electric.
HJC	Holmquist-Johnson-Cook.
IN718	Inconel 718.
ISO	International Organization for Standardization.
JC	Johnson-Cook.
LPBF	laser powder-bed fusion.
MATLAB	Matrix Laboratory.
MTS	MTS Systems Corporation.
NURBS	Non-Uniform Rational B-Splines.
PBF	powder-bed fusion.
SEM	scanning electron microscopy.
SHPB	Split Hopkinson Pressure Bar.
SIMP	Solid/Isotropic Material and Penetration.

1 Introduction

The primary objectives of this research into the time-dependent characteristics of lattice structures were threefold. First was to characterize the quasi-static and dynamic material properties of lattice-designed structures. Second was to develop a constitutive model for the various lattice structures to determine their material response. Third was to create a computational model for the simulation of impact under multiple design considerations and constraints.

1.1 OVERVIEW

Advances in new materials, manufacturing techniques, and structural designs have the potential to change how engineers view and address problems drastically. Imagine a highly porous metal cellular structure that is significantly lighter than its solid counterpart but provides improved energy absorption and dispersion characteristics while maintaining an advantageous strength-to-weight ratio. The implications of this cellular design to personnel and vehicle survivability and ballistic design are enormous. A significant effort within this research was to compare the high strain rate impact performance of advanced structural designs incorporating lattice architectures. This effort required the evaluation of the material properties of various metal periodic cellular structures, examining how the relative density and cell design affected the global material properties. Additive manufacturing (AM) techniques were used in the fabrication of the designed cellular structures. These properties were compared with the baseline solid material to determine the change in specific material properties, or the ratio of the property in question to the design relative density, along with a comparison between the structural designs of energy absorption. The ultimate goal was to determine if the use of these advanced designs improved impact performance in a high strain rate application. The remainder of this chapter provides an outline of research objectives, background information on critical topics, and a discussion of how this textbook is organized.

1.2 RESEARCH OBJECTIVES

The overarching goal of this research was to better understand the time-dependent characteristics of cellular structural designs and how those characteristics affected dynamic failure performance under high strain rate impacts. Several intermediate research objectives were determined to reach this goal and are listed below as objective statements, along with the initial hypothesis of the objective's efficacy and necessary research efforts to achieve the objectives.

DOI: 10.1201/9781003299639-1

1.2.1   Research Objective 1: Characterize Quasi-Static Material Properties of Cellular Designs

Baseline equivalent static materials properties had to be determined for each of the internal lattice designs proposed. The properties of interest were the modulus of elasticity, yield strength, plateau stress, densification strain, toughness, and Poisson's ratio. These parameters along with the relative density of the manufactured specimens were used to determine specific material properties such as specific yield strength and specific toughness.

Hypothesis 1: Laboratory compression testing will provide the data necessary to determine equivalent static material properties for each of the designs, such that the lattice designs can be down-selected for subsequent experimentation.

Research Effort 1: Laboratory compression testing of each cellular design was performed, and the specific material properties of each design were compared to determine if designs were suitable for further evaluation.

1.2.2   Research Objective 2: Characterize Dynamic Material Properties of Cellular Designs

Baseline equivalent dynamic material properties had to be determined for each of the internal lattice designs chosen from the static testing results. Equivalent plasticity and damage modeling coefficients were required for inclusion in the computational ballistic impact model.

Hypothesis 2: The Taylor impact test will provide all the data necessary to determine equivalent damage modeling coefficients. This hypothesis is based on the work done by Johnson and Holmquist, *Evaluation of Cylinder-Impact Test Data for Constitutive Model Constants* [3].

Research Effort 2: Taylor impact tests were performed at various impact velocities with the chosen lattice designs to determine an equivalent flow stress model and damage modeling parameters. Split Hopkinson Pressure Bar testing was also performed to supplement the data set used to determine the lattice dynamic material properties.

1.2.3   Research Objective 3: Develop Computational Model and Impact Simulation

Computational modeling and simulation were required to evaluate a high-rate projectile impact with solid and brittle targets. This model served as the baseline for comparison against modified projectiles, incorporating cellular structures, and therefore needed to be validated against experimental data.

Hypothesis 3: A validated physics-based computational impact model using finite element analysis (FEA) techniques will provide the necessary means for the comparison of projectile designs.

Research Effort 3.A: A computational model of the high strain rate impact of a baseline projectile against a target was developed.

Research Effort 3.B: The developed model was validated against experimental data sets.

1.2.4 RESEARCH OBJECTIVE 4: EVALUATE INCLUSION OF CELLULAR STRUCTURES ON IMPACT RESULTS

The cellular structures were incorporated into the impact model for simulation, with a focus on the inclusion of the lattice design into the projectile nose. The equivalent material properties for each of the cellular designs found during experimentation were incorporated within the model. Primary performance measures included projectile depth of penetration, stress wave propagation, energy absorption and control, and projectile failure patterns.

Hypothesis 4: The impact model will provide the means for performance comparison between the structural designs and applications, considering predicted penetration depth, stress wave absorption and control, and deformation patterns of both projectile and target.

Research Effort 4: Impact simulations were performed using AM projectiles that incorporated a cellular architecture section under the same initial and boundary conditions to determine changes in performance between the designs.

1.3 BACKGROUND

As mentioned, this research was primarily concerned with and focused on strain rate sensitivity of novel lattice structures. One of the best means to test and compare their dynamic response was to place the lattice within a projectile impact environment. The nature of this research requires the introduction and discussion of several intersecting engineering disciplines. The background begins with a discussion of cellular structure designs in Section 1.3.1, which describes how material properties are characterized for cellular architectures, going into further examination of lattice and triply periodic minimal surface (TPMS) designs. Then a brief overview of AM methods is presented in Section 1.3.2, which will be used extensively during this research. Finally, Section 1.3.3 introduces the methods used for the modeling and simulation of impact testing.

1.3.1 CELLULAR STRUCTURES

This section establishes the background and terminology regarding cellular structures and architecture, and then expands in further detail on lattice designs along with the subset of lattices, TPMS. While the use of natural cellular materials, such as wood and cork, has been around for centuries, designed cellular architectures represent a relatively new classification of engineering material [4]. Cellular materials and designs are characterized by a combination of solid material and gas, typically air. The significance of cellular structures is that they provide high strength and stiffness with low density while often affording superior energy absorption characteristics, which is why they were chosen for analysis under high strain rate considerations.

Three primary factors dictate the resultant material properties of a cellular structure. They are the properties of the material from which the structure is made, the topology or structural design used, and the relative density of the cellular design [5].

1.3.1.1 Terminology

The field of cellular technologies is quite different from that of continuum technologies; therefore, an introduction of specific features must be defined, and relevant terminology must be presented.

- **Relative density**: Arguably, the most important characteristic of a cellular design is its relative density, which is the density of the cellular structure divided by the density of the material from which it was constructed; see Equation 1.1.

$$rd = \frac{\rho^*}{\rho_s}$$

(1.1)

  In this equation, ρ^* is the volumetric mass density, which is referred to as density moving forward, of the cellular structure, and ρ_s is the density of the base material. As noted by Gibson and Ashby, when the relative density of a designed cellular structure increases above approximately 0.3, the material properties reflect those of a solid material containing pores rather than a cellular structure [6].
- **Porosity**: Closely related to relative density is porosity or the proportion of void space within the structure. It is often used interchangeably with the relative density, as it is the complement to the relative density, given by Equation 1.2.

$$p = \left(1 - \frac{\rho^*}{\rho_s}\right)$$

(1.2)

- **Cell size**: Sometimes referred to as the cell length or mean cell diameter, the cell size is the average cell size measured as either the cross-sectional distance or diameter.
- **Cell wall thickness**: Cell wall thickness, often referred to simply as thickness, is the average thickness of the cell edges or structural members of the cell.

1.3.1.2 Classification

Multiple different types of classification can be used to describe cellular structures, typically related to cellular distribution or design. A couple of these classification methods will be explained in detail.

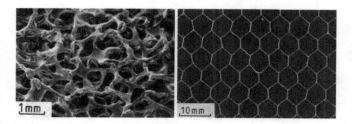

FIGURE 1.1   Non-periodic versus periodic cellular architectures [6].

The first method of classification to be discussed is periodic versus non-periodic distribution. A periodic cellular architecture is defined by having a unit cell design replicated to fill a structural space. In contrast, a non-periodic cellular architecture is defined by the void space within a structure being distributed in a random or stochastic fashion. Figure 1.1 shows the difference between periodic and non-periodic distributions.

Cellular architectures can also be split into two categories based on cellular design: closed cell or open cell. Sealed voids within the structure characterize closed cell architectures. In comparison, open cell architectures are described by an interconnected network of gas-filled space within the structure. Figure 1.2 shows the difference between open cell and closed cell structures in metal foams.

1.3.1.3   Mechanical Properties

As previously mentioned, the use of natural cellular materials has been around for centuries; however, an understanding of their material properties and how they are related to their design characteristics did not truly emerge until the late 1950s with the work of Gent and Thomas [8]. Their initial work primarily investigated the physical properties of non-periodic polymer foams. Patel and Finnie continued this work in the 1960s and 1970s, which expanded the research into evaluating the thermal properties of non-periodic polymer foams [9]. The efforts of these four individuals laid the groundwork for what can be considered one of the foundational texts on cellular materials, *Cellular Solids: Structure and Properties* by Gibson and Ashby, first published in 1988 [6]. This work primarily considers natural cellular materials, but manufactured cellular materials, such as metal foams and honeycombs, are included. It is worth noting that much of the analysis outlined in *Cellular Solids* can be applied as the basis of analysis for periodic cellular structures. However, when evaluating the mechanical properties of periodic cellular architectures, there are distinct differences in the compression response of strut-based and surface-based lattices.

Lattice structural designs are a specific subset of cellular architectures defined by periodic replication, or symmetry, of individual cellular units with well-defined structural characteristics, typically incorporating an open cell design. Therefore, lattices are often characterized as periodic and open cell structures. Two common variations of lattice structures are strut-based and surface-based networks [10]. Strut-based networks can be considered stretch-dominated structures and are defined by a joint frame design similar to a truss [11]. As with the truss, these networks have structural members that are loaded through tension or compression, leading to members stretching to carry a load. See Figure 1.3 for an example of a strut-based cell and structure.

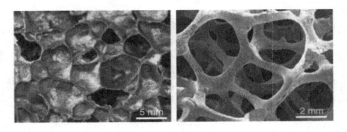

FIGURE 1.2   Closed cell versus open cell architectures [7].

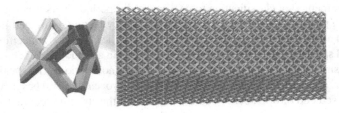

FIGURE 1.3   Strut-based lattice cell and structure.

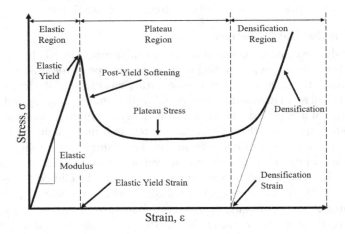

FIGURE 1.4   Typical strut-based lattice stress–strain response.

Figure 1.4 depicts a typical stress–strain curve for a strut-based lattice design under compressive loading. This curve represents three different phases in the lattice response. The first phase of the response is linear-elastic. In this region, the modulus of elasticity (E), or Young's modulus, of the design can be determined. Next, the material exhibits post-yield softening before reaching a plateau. This softening is due to the strut-based failure modes of tension and compression, namely buckling or brittle failure. The plateau region of the response curve is due to strut collapse. Because of the brittle nature of the collapse of the struts, the plateau region is shorter and quickly gives way to the third phase. The third and final stage in the design response is called densification, which is noted by a sharp rise in stress. Here, all the cells within the structure have collapsed onto one another, and the material is near solid again.

Surface-based networks can be thought of as bending-dominated structures and respond in a similar fashion to foams [11]. While described as bending-dominated, these networks can see failure through bending, buckling, or crushing. The common trend is that structural failure happens through the thickness of the surface. Figure 1.5 provides an example of a surfaced-based lattice cell and structure. Figure 1.6 depicts a typical stress–strain curve from compression testing of a surface- based lattice design. Again, there are three distinct phases within this curve. First, similar to the strut-based lattice, the surface-based design displays a linear-elastic response up to its elastic yield strength. Surface-based lattices will generally have a lower yield point than a strut-based design of similar relative density due to their bending failure modes. The second phase of the response curve is the plastic response of the cells, characterized by plateau stress (σ_{pl}). The final stage of the stress–strain curve is once again densification.

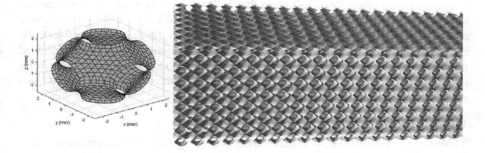

FIGURE 1.5 Surface-based lattice cell and structure.

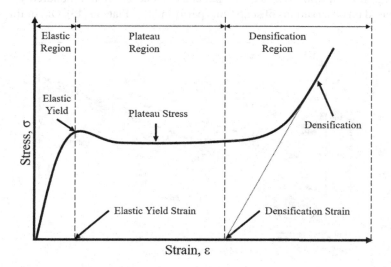

FIGURE 1.6 Typical surface-based lattice stress–strain response.

These differences in the mechanical behavior indicate that each structure may be better suited for different applications. The nature of loading and failure of the strut-based networks suggest that these architectures would perform better under uniaxial loading than the surface-based designs and yield a higher strength-to-weight ratio.

However, the plateau region of the surface-based structure response tends to be longer than that of an equivalent strut-based design, which indicates that the surfaced-based design would perform better in energy-absorbing applications [5]. Because this research is, in part, interested in the energy absorption and dispersion character-istics of lattice structures, the primary focus will be on surface-based structures. The relative strength versus relative density for different cellular structure types and topologies can be seen in Figure 1.7. It can be seen that, under normal circumstances, strut-based lattices have higher relative strength values than their surfaced-based counterparts, but the relative strength gained by an incremental change in relative density is greater for the surface-based structures. Because of the limited research on the relative strength of the periodic lattices being evaluated in this research, it is unknown where precisely on this chart they will fall. However, the relative densities being considered as part of this research fall within the range of 8%–40% relative density, which narrows the region of interest to where the difference between the relative strength of strut-based and surface-based structures is minor. As mentioned previously, the energy absorption characteristics of the structures under high strain rate impact are of greater interest than the relative compressive strength, so the focus will be primarily on surfaced-based periodic lattices.

1.3.1.4   Triply Periodic Minimal Surface (TPMS)

TPMSs are a classification of surface-based lattices. Research into minimal surfaces began with the work of Lagrange in the mid-1700s through his analysis of varia-tional problems and minimizing surface area [12]. The first mathematical proposi-tions regarding minimal surfaces did not come about for another hundred years and were based on observations of soap films performed by Plateau [13]. Out of this work

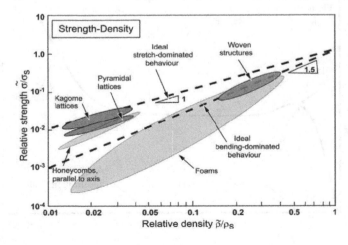

FIGURE 1.7   Relative strength versus relative density of cellular structures [5].

came the *Plateau Problem*, a mathematical problem based on the finding the minimal surface spanned by a given curve, which gave way to the research into infinitely periodic minimal surfaces, or in the natural world TPMS [14]. Schwarz described the first TPMS designs in 1866 [15]; this work continued with his students, all together discovering five TPMS [16]. One of his most gifted students, Neovius, continued the study of minimal surfaces in the 1880s, discovering his own TPMS [17]. These lattices are unique structured surfaces that are described by three distinct criteria [18]. First, the surfaces are symmetric and periodic in all three axes. Second, the surfaces are area-minimizing. That is, the surface comprises the smallest possible area that bounds the region. Finally, the surfaces have zero mean curvature over a single cell or symmetric structure. The periodic nature of these designs allows them to be patterned to fill specific dimensional spaces.

Several different TPMS designs have been found and are well described in Schoen's seminal work on periodic surfaces [18]. Four of these designs were chosen for initial analysis and will be described in more detail. These lattice structures are the primitive (Schwarz P), diamond (Schwarz D), Schoen's I-WP, and Lidinoid. In the lattice surface equations shown below, x, y, and z are the surface's Cartesian coordinates in three-dimensional (3D) space, and m represents the periodicity scaling factor, which is the ratio of cell size to π, that sets the cell density throughout the structure.

The primitive lattice structure was chosen due to its geometry. The transition networks between cells have circular cross-sectional areas, which would limit any stress concentration points across the surface. They would also provide a more uniform stress distribution throughout the structure. The primitive surface can be approximated by Equation 1.3 [19].

$$P : \cos mx + \cos my + \cos mz = 0 \tag{1.3}$$

A depiction of a primitive surface cell is presented in Figure 1.8.

The diamond and I-WP surface-based lattice structures were chosen due to the initial compressive strength and toughness testing accomplished by Al-Ketan et al. [20]. Their results indicated the highest plateau stresses and toughness values for these two designs, with the diamond structure exhibiting the best mechanical properties. The diamond and I-WP surfaces can be approximated by Equations 1.4 and 1.5, respectively [19].

$$D : \sin mx \sin my \sin mz + \sin mx \cos my \cos mz$$
$$+ \cos mx \sin my \cos mz + \cos mx \cos my \sin mz = 0 \tag{1.4}$$

$$\text{IWP} : 2(\cos mx \cos my + \cos mx \cos mz + \cos my \cos mz)$$
$$- \cos 2mx \cos 2my \cos 2mz = 0 \tag{1.5}$$

A depiction of a diamond surface cell is presented in Figure 1.9.

A depiction of an I-WP surface cell is presented in Figure 1.10.

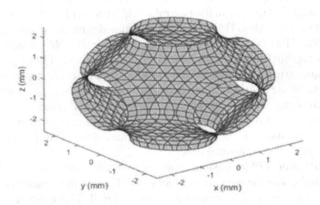

FIGURE 1.8   Primitive surface cell.

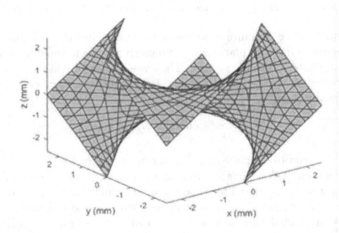

FIGURE 1.9   Diamond surface cell.

The Lidinoid surface-based lattice structure was chosen due to the lack of information available on its mechanical properties and its intriguing geometry. The Lidinoid surface can be approximated by Equation 1.6 [19].

$$L : \frac{1}{2}\left(\sin 2mx \cos my \sin mz + \sin 2my \cos mz \sin mx + \sin 2mz \cos mx \sin my\right)$$

$$-\frac{1}{2}\left(\cos 2mx \cos 2my + \cos 2my \cos 2mz = \cos 2mx \cos 2mz\right) + 0.15 = 0 \tag{1.6}$$

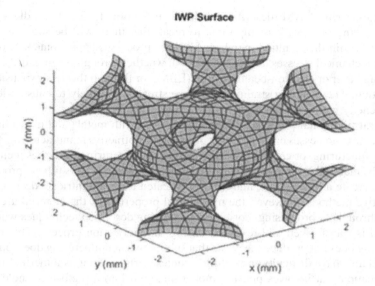

FIGURE 1.10   I-WP surface cell.

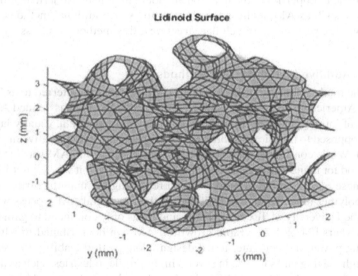

FIGURE 1.11   Lidinoid surface cell.

A depiction of a Lidinoid surface cell is presented in Figure 1.11.

1.3.2   MANUFACTURING METHODS

This section introduces the background and terminology associated with material manufacturing or fabrication methods. Traditional manufacturing methods will be briefly introduced to help form the basis for understanding advanced manufacturing techniques, such as AM, described in detail in Section 1.3.2.1. Regardless of the manufacturing process, it is important to recall that there will be some degree of disparity within the resultant member. All these processes apply some sort of thermal or mechanical stresses during fabrication, whether through forging, die-casting, extruding, laser or electron-beam melting (EBM), or through thermal expansion and contraction. The resulting strains from these stresses are likely to cause a level of heterogeneity within the part.

Several traditional processing methods are used with metals and their alloys; in general, these processes can be thought of as either subtractive manufacturing, formative manufacturing, or casting [21]. Subtractive manufacturing describes techniques where the material is removed during the fabrication process. This type of processing tends to create a lot of waste, and it is complicated to create intricate designs using subtractive methods. However, the mechanical properties of the material are maintained throughout processing. Formative processing described techniques where the material is transformed and reshaped during the fabrication process. This type of processing does not produce the waste that subtractive manufacturing does, but these methods also have difficulty producing complex parts. Casting is a thermal method of fabrication that involves pouring molten material into a preformed mold of the desired shape. This technique allows for the fabrication of these complex parts, but the mechanical properties of the material are not maintained. Modern manufacturing techniques, such as AM, address the shortcomings of traditional methods, and due to the complex geometries of cellular structures, this method is necessary for their fabrication.

1.3.2.1 Additive Manufacturing Methods

One of the modern fabrication techniques is AM, sometimes referred to as 3D printing. The American Society for Testing and Materials (ASTM) has defined AM as the "process of joining materials to make parts from 3D model data, usually layer upon layer, as opposed to traditional subtractive manufacturing and formative manufacturing" [22]. While considered modern, the first patent regarding AM was filed in 1892 as a method for manufacturing contour relief maps [23]. Yet, it was not until 1981 that the Japanese published the first information regarding the manufacturing of a solid printed polymer model, and then in 1986 that the first metal AM process was developed at the University of Texas at Austin, and it has only continued to gain attention in recent years [24–26]. The rapid growth in AM can be attributed to advances in computing power and computer-aided design, along with the ability to create complex, weight-saving parts for use in several high- profile industries, such as aerospace and automotive. Advances in AM technology have led to decreased monetary and time costs for part fabrication, putting it on the same level or ahead of some traditional manufacturing techniques for commercial use. Committee F42 on Additive Manufacturing Technologies from ASTM, working alongside the International Organization for Standardization (ISO) Technical Committee 261 for Additive Manufacturing, released ISO/ASTM 52900 in 2015 to standardize the field of AM [22]. The goals of ISO/ASTM 52900 were to define the AM processes, develop standardized definitions and terminology, as well as provide guidance on test procedures and measures of quality, some of which are also contained in ISO/ASTM 52921

[27]. The ISO/ASTM standards define seven AM methods for use in manufacturing; four of these techniques can be used to fabricate metal parts. Some of the more prevalent techniques are direct energy deposition and powder-bed fusion (PBF). PBF is defined by ISO/ASTM 52900 as the "process in which thermal energy selectively fuses regions of a powder bed" [22]. PBF uses a high-power energy source, such as a laser, electron beam, or plasma arc, to melt the raw material. The process name can be modified depending on the energy source used, such as EBM and laser powder-bed fusion (LPBF). Regardless of the energy source, the process for fabrication is nearly identical. LPBF was the method used to fabricate the cellular structures for experimental testing and will be described in more detail below.

In LPBF fabrication, the printer applies a thin layer of powder across the build surface. The laser then melts and fuses the powder based on a computed scan strategy that determines the build pattern and timing information. Once a layer is completed, the printer adds another layer of powder, and the process is repeated until the build is complete [27]. There are five primary parameters that can be controlled during the print process: laser power, scan speed, scan spacing, scan pattern, and powder layer thickness. Laser power, scan speed, and scan spacing are closely linked parameters that must be balanced to achieve an optimal build. The lower the laser power, the slower the scan speed must be to produce the same melt pool; if the melt pool size is changed, the scan spacing must be altered to ensure adequate overlap for fusing within the build. Jia and Gu evaluated the impact of the laser power and scan speed on the mechanical properties of AM Inconel 718 (IN718). They found significant variation in densification and microstructure, both of which affect the resultant material properties of the manufactured parts [29]. In addition, the scan strategy used to build the part can be modified, with either a single or multiple strategy concept used within a build layer. A typical build begins with a contour scan to define the part boundaries, and then the part outline is filled using a raster scan strategy. Three of the most common raster-fill scan strategies are continuous, island, and strip. The scan strategies used can result in microstructure changes due to heating levels and melt pool sizes at particular build locations [30], as well as the thermophysical properties of the alloy being printed. The scan path taken can lead to part distortion, which can lead to the accumulation of residual stresses within a part [31]. A pictorial representation of these strategies is shown in Figure 1.12.

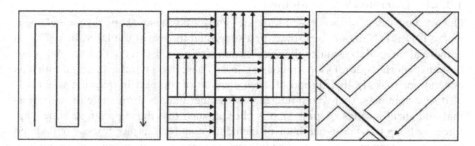

FIGURE 1.12 Powder-bed fusion scan strategies: continuous, island, and strip [26].

Just as the other parameters can affect the build properties, the powder layer height can also be adjusted to modify the melt pool, which will, in turn, affect the build precision and part microstructure [32].

A common concern when using AM is that the build direction will cause anisotropy in the mechanical properties of the manufactured part. Al-Ketan et al. performed compressive testing of steel lattice samples, tested both parallel and perpendicular to the print direction, and found that the mechanical response was very similar for both test directions [10]. This indicates that there is no noticeable anisotropic response due to the AM process for lattice structures.

1.3.3 IMPACT MODELING AND SIMULATION

This research intends to evaluate the strain rate sensitivity of periodic lattice designs, which will be accomplished within a projectile impact environment. Some foundational information will be presented on impact modeling and simulation. This section will provide contextual background information, introduce relevant terminology, and highlight some of the methods used in the study of high strain rate impacts.

High strain rate impact testing entails the high-velocity collision of a projectile into a target and observing the effects of the interaction. This analysis is done in terms of deformation and fragmentation of both the projectile and target, and it evaluates the target failure pattern and projectile penetration depth [33]. Due to the expense of these impact experiments, especially considering hyper-velocity ballistic applications, much of this study is done through numerical and computational modeling and simulation [34]. The modeling and simulation of high strain rate impact is complicated, involving the evaluation of several non-linear steps, such as contact between the projectile and target, high strain rates near the impact region, stress wave propagation through both the projectile and the target, and material deformation and separation. Due to these sources of non-linearity, model validation is essential. A thorough impact evaluation provides the validation of numerical failure models through the assessment of the energy absorbed during impact, penetration depth in a thick target or residual velocity of the projectile in the thin target, and the measured deformation of both projectile and target [35]. Several methods have been devised to provide insight into the impact event; two of these methods will be explained in further detail below.

1.3.3.1 Discrete Element Method

One of the earliest methods used to model impact physics was the discrete element method (DEM). Cundall and Strack first suggested the use of the DEM in 1979 as a numerical method to evaluate the stresses and strains within a continuum of discontinuous materials [36]. Discontinuous materials describe granular materials such as concrete or rock. In their proposal, each element of the part in question was modeled as a rigid mass that would interface with other masses through deformable normal and shear springs. A time step is chosen, such that during a single time step, forces will not transfer to neighboring elements, and the interface forces are updated between subsequent time steps. The interactions are solved using an explicit numerical method to evaluate equations of motion for each discrete element. Sitharam first demonstrated that this method could successfully predict the results of compression testing and three-point bend testing of concrete [37]. Nair and Rao further applied the principles of DEM to ballistic impact [34]. They developed a one-dimensional ballistic impact simulation based on Cundall and Strack's equations to describe disk compression between rigid walls. The one-dimensional simulation performs well at slower impact velocities, less than 200 m/s, but it does not accurately portray the projectile penetration depth at higher velocities [34]. While providing promising results, numerical DEMs have not been applied to 2D or 3D simulations, primarily due to computational expense.

1.3.3.2 Finite Element Method

One of the more prevalent methods used to simulate and model impacts is the finite element method (FEM). FEM, sometimes also referred to as FEA, is a numerical method used to find the solution to boundary value problems. FEM grew out of the aerospace industry in the 1960s as a method for performing stress and thermal analysis on complex aircraft, rockets, and engine parts [38]. FEM uses the concept of discretization to divide a body into smaller units, finite elements, that are interconnected across shared nodes or boundaries. Applicable field quantities are approximated across the entire structure through piecewise element interpolation and summation [39]. Gailly and Espinosa used FEM to describe the failure modes of brittle materials during ballistic penetration, finding where the transition from micro-cracking to pulverization occurs in ceramic armor [40]. Using FEM, Gailly and Espinosa created a two-dimensional impact model that correlated well with experimental results even at higher impact velocities, up to 1,450 m/s. FEM can also provide accurate 3D simulations at increased velocities, as demonstrated by Kurtaran et al. in 2003 [41]. In this study, Kurtaran et al. were able to model the deformation patterns of a projectile and target up to 1,500 m/s, as well as show the effects of thermal softening in target deformation at higher velocities. As demonstrated by the studies mentioned above, one benefit of FEM over DEM is the relative computational ease to perform simulations in 2D and 3D while maintaining a good correlation with experimental and real-world results.

Due to the nature of evaluating cellular designs under high strain rate impact, FEA was used to model the interactions during the impact event. Fundamental FEA theories are discussed in detail in Chapter 2.

1.4 ORGANIZATION OF SUBSEQUENT CHAPTERS

The following chapters expand on the contents presented in this first chapter.

Chapter 2 provides the theoretical context used as a foundation for the research and analysis in this study. A further examination of FEM was performed to detail traditional Lagrangian explicit analysis methods in Section 2.1.1. Section 2.1.2 describes the FEM of smoothed particle hydrodynamics (SPH) as it applies to high strain rate modeling. A discussion on shock waves due to impact and the subsequent use of an equation of state is presented in Section 2.1.3. Chapter 2 also provides an in-depth discussion of damage modeling theory, going into further detail on the Johnson–Cook (JC) and Holmquist–Johnson–Cook failure models used as part of this research, in Section 2.2. Finally, Section 2.3 presents the Rusch and Gibson modified constitutive model theories regarding cellular solids and structures.

Chapter 3 introduces the experimental and computational tools used to acquire and analyze the data required for this research. Section 3.2 discusses the methods used to fabricate test specimens and then further outlines the experimental techniques used to determine cellular structure material properties. This section goes into further detail to explain the methods incorporated for both the quasi-static regime and dynamic property regime. Computational methods used in the modeling and simulation of ballistic impact are discussed in Section 3.3, building on the FEM theory presented in Section 2.1.

Chapters 4–6 present the results and discussion of the experimental and computation testing performed as part of this research. These chapters are laid out to follow the intermediate objectives determined as the path for evaluating the time-dependent characteristics of the lattice structures.

Chapter 4 starts with the initial evaluation of the lattice designs tested within various specimen configurations. The results from these evaluations provided the framework for all further quasi-static uniaxial compression testing.

Chapter 5 provides the experimental results that characterize the quasi-static mechanical properties of the lattice structures, as well as evaluating variational designs of the base lattice cells.

Chapter 6 presents the results from the Split Hopkinson Pressure Bar testing, evaluating the change in material properties due to an increase in applied strain rate. The JC plasticity model was assessed over the tested strain rate range, with further discussion presented on the utility of this flow stress model for lattice structures.

Chapter 7 provides the results from the Taylor impact testing that was performed using the lattice structures and develops a new strain rate and temperature-sensitive flow stress model to be used with lattice architectures. Chapter 8 evaluates the use of the JC Damage Model for lattices under high strain rate compression.

Chapter 9 introduces the FEA impact modeling techniques as part of a simulation study that compares the traditional grid-based Lagrangian methods with those of SPH.

Chapter 10 compares the final impact experimentation of a projectile incorporating a lattice section against its computational counterpart. Penetration testing performance measures, along with computational stress wave analysis, are used as means of comparison.

Chapter 11 presents the conclusions.

REFERENCES

3. G. R. Johnson and T. J. Holmquist. Evaluation of cylinder-impact test data for constitutive model constants. *Journal of Applied Physics*, 64(8): 3901–3910, 1988.
4. L. J. Gibson. Cork: Structure, properties, applications. *Arnoldia*, 74(1): 23–27, 2016.
5. M. Scheffler and P. Colombo, editors. *Cellular Ceramics*. Wiley-VCH, Hoboken, New Jersey, 2005.
6. L. J. Gibson and M. F. Ashby. *Cellular Solids: Structure and Properties*. Cambridge University Press, Cambridge, England, 2nd edition, 1997.

8. A. N. Gent and A. G. Thomas. Failure of foamed elastic materials. *Journal of Applied Polymer Science*, 2(6): 354–357, 1959.
9. M. R. Patel and I. Finnie. Structural features and mechanical properties of rigid cellular plastics. *Journal of Materials*, 5: 909–932, 1970.
10. O. Al-Ketan, R. Rowshan, A. N. Palazotto, and R. K. Abu Al-Rub. On mechanical properties of cellular steel solids with shell-like periodic architectures fabricated by selective laser sintering. *Journal of Engineering Materials and Technology, Transactions of the ASME*, 141(2): 1–12, 2019.
11. M. F. Ashby. *Materials Selection in Mechanical Design*. Elsevier, Amsterdam, Netherlands, 3rd edition, 2005.
12. J. Lagrange. *Oeuvres*. Gauthier-Villars, Paris, 1867.
13. J. Plateau. *Statique experimental et theoretique des liquides soumis aux seule forces moleculair*. Gauthier-Villars, Paris, 1873.
14. H. Karcher and K. Polthier. Construction of triply periodic minimal surfaces. *Philosophical Transactions: Mathematical, Physical and Engineering Sciences*, 354(1715): 2077–2104, 1996.
15. H. A. Schwarz. *Bestimmung einer speziellen Minimalfl¨ache*. Berlin Academy, Berlin, Germany, 1871.
16. H. A. Schwarz. Bestimmung einer speciellen Minimalfl¨ache. In *Gesammelte Mathematische Abhandlungen*, pages 6–91. Springer Berlin Heidelberg, Berlin, Heidelberg, 1890.
17. E. R. Neovius. *Bestimmung zweier spezieller periodischer Minimalflächen*. Helsingfors, Finland, 1883.
18. A. H. Schoen. *Infinite Periodic Minimal Surfaces without Self-Intersections*. NASA, tn d-5541 edition, 1970.
19. S. N. Krivoshapko and V. N. Ivanov. Minimal surfaces. In *Encyclopedia of Analytical Surfaces*, pages 427–457. Springer International Publishing, 2015.
20. O. Al-Ketan, R. Rowshan, and R. K. Abu Al-Rub. Topology- mechanical property relationship of 3D printed strut, skeletal, and sheet based periodic metallic cellular materials. *Additive Manufacturing*, 19(January): 167–183, 2018.
21. J. T. Black and R. A. Kohser. *DeGarmo's Materials and Processes in Manufacturing*. John Wiley & Sons, Inc., Amsterdam, Netherlands, 11th edition, 2012.
22. ASTM International. Standard terminology for additive manufacturing – General principles – Terminology, 2015.
23. J. E. Blanther. Manufacture of contour relief maps, Patent, US, 1892.
24. C. R. Deckard. Part generation by layer-wise selective laser sintering. Technical report, University of Texas at Austin, 1986.
25. A. Lou and C. Grosvenor. *Selective Laser Sintering, Birth of an Industry*. Department of Mechanical Engineering, University of Texas, Austin, TX, 2012.
26. W. J. Sames, F. A. List, S. Pannala, R. R. Dehoff, and S. S. Babu. The metallurgy and processing science of metal additive manufacturing. *International Materials Reviews*, 61(5): 315–360, 2016.
27. ASTM International. Standard terminology for additive manufacturing — Coordinate systems and test methodologies, 2013.
29. Q. Jia and D. Gu. Selective laser melting additive manufacturing of Inconel 718 superalloy parts: Densification, microstructure and properties. *Journal of Alloys and Compounds*, 585: 713–721, 2014.
30. X. Wang, X. Gong, and K. Chou. Review on powder-bed laser additive manufacturing of Inconel 718 parts. *Proceedings of the Institution of Mechanical Engineers, Part B: Journal of Engineering Manufacture*, 231(11): 1890–1903, 2017.
31. Ian Gibson, David Rosen, and Brent Stucker. *Additive Manufacturing Technologies*. Springer New York, New York, NY, 2015.

32. V. S. Sufiiarov, A. A. Popovich, E. V. Borisov, I. A. Polozov, D. V. Masaylo, and A. V. Orlov. The effect of layer thickness at selective laser melting. *Procedia Engineering*, 174: 126–134, 2017.

33. G. Ben-Dor, A. Dubinsky, and T. Elperin. Ballistic impact: Recent advances in analytical modeling of plate penetration dynamics-a review. *Applied Mechanics Reviews*, 58(1–6): 355–370, 2005.

34. R. P. Nair and C. Lakshmana Rao. Simulation of depth of penetration during ballistic impact on thick targets using a one-dimensional discrete element model. *Sadhana - Academy Proceedings in Engineering Sciences*, 37(2): 261–279, 2012.

35. A. D. Resnyansky and G. Katselis. *Ballistic and Material Testing Procedures and Test Results for Composite Samples for the TIGER Helicopter Vulnerability Project*. Australian Government, Department of Defense, dsto-tr-16 edition, 2004.

36. P. A. Cundall and O. D. L. Strack. A discrete numerical model for granular assemblies. *G´eotechnique*, 29(1): 47–65, 1979.

37. T. G. Sitharam. Numerical simulation of particulate materials using discrete element modelling. *Current Science*, 78(7): 876–886, 2000.

38. Edward L. Wilson. Structural analysis of axisymmetric solids. *AIAA Journal*, 3(12):2269–2274, 1965.

39. R. D. Cook, D. S. Malkus, M. E. Plesha, and R. J. Witt. *Concepts and Applications of Finite Element Analysis*. John Wiley & Sons, Inc., Hoboken, New Jersey, 2002.

40. B. A. Gailly and H. D. Espinosa. Modelling of failure mode transition in ballistic penetration with a continuum model describing microcracking and flow of pulverized media. *International Journal for Numerical Methods in Engineering*, 54(3): 365–398, 2002.

41. H. Kurtaran, M. Buyuk, and A. Eskandarian. Ballistic impact simulation of a model vehicle door using finite element method. *Theoretical and Applied Fracture Mechanics*, 40(2): 113–121, 2003.

2 Background Theory

The scope of the research contained herein focuses on the damage response and failure mechanisms of advanced structural designs, including triply periodic minimal surface structures, under high strain rate impact. The nature of this research requires the introduction and discussion of several intersecting engineering disciplines. This chapter provides the necessary foundational theory to support the experimental methodology and analysis required for this research.

2.1 FINITE ELEMENT METHOD

This research entails modeling a dynamic problem that incorporates large strains, high strain rates, and rate-dependent material properties. These characteristics drive the necessary selection of finite element method (FEM) modeling approach and features, explained in the following sections. Since Dassault Syst`emes Abaqus finite element software will be used as the primary means of computational analysis for this research, the specific applications within Abaqus will be highlighted throughout this chapter.

2.1.1 FINITE ELEMENT METHOD THEORY

FEM, or finite element analysis (FEA), is a numerical method for solving complex boundary value problems where analytical solutions may not be able to be obtained. Two approaches to be considered when solving a finite element problem are the implicit approach and the explicit approach. The implicit approach is most useful when considering static problems or modal analysis, where time dependence of the solution is not essential. It follows the conventional finite formulation that includes formulating a global stiffness matrix and global mass matrix that forces global equilibrium at every time step. Whereas the explicit approach is most useful in solving problems that result in large deformations or are highly time dependent. Explicit analysis will be used in this research since it deals with the time-dependent behavior of cellular structures and the high strain rate effects of impact.

The explicit analysis approach to account for the deformation rate and stress wave propagation of impact must consider displacement, velocity, and acceleration on a node-by-node basis within the structure with respect to time. Thus, it relies on a finite difference scheme incorporating central difference time integration (CDTI) to calculate nodal field variables as the time step is incremented along with the use of lumped mass matrices. First, the CDTI satisfies equilibrium of the equations of motion at time t, then accelerations at time t are used to determine the velocity at time $t + \Delta t/2$, which is then, in turn, used to determine the displacement at time $t + \Delta t$. This process is repeated throughout the time interval of interest and evaluates deformation and stress wave propagation through the part. In CDTI, the expressions

DOI: 10.1201/9781003299639-2

for displacement (u), velocity ($\dot{u}$), and acceleration ($\ddot{u}$) are given by Equations 2.1–2.3, respectively [42].

$$u_{(i+1)} = u_{(i)} + \Delta t_{(i+1)} \frac{du}{dt}_{\left(t+\frac{1}{2}\right)} \tag{2.1}$$

$$\dot{u}_{\left(i+\frac{1}{2}\right)} = \dot{u}_{\left(i-\frac{1}{2}\right)} + \frac{\Delta t_{(i+1)} + \Delta t_{(i)}}{2} \ddot{u}_{(i)} \tag{2.2}$$

$$\ddot{u}_{(i)} = M^{-1}\left(F_{(i)} - P_{(i)}\right) \tag{2.3}$$

In these equations, the subscript i is the time increment number, and therefor $t - \frac{1}{2}$ and $t + \frac{1}{2}$ refer to the mi-increment values, Δt refers to the time step, M is lumped mass matrix, F is the applied external load vector, and P is the internal force vector [43]. The internal force vector is determined through an analysis of the element stress–strain relations.

In FEM, elemental strain, ε, is determined from displacement through the utilization of the strain-displacement matrix, shown in Equation 2.4.

$$\varepsilon = Bu_{(i)} \tag{2.4}$$

The strain-displacement matrix, B, relates the displacements to the element strain component based on the derivatives of the element shape functions or basis functions. The strain determined from Equation 2.4 is the total strain, which is comprised of both the elastic and viscoplastic strain. Elastic strain is the fully recoverable deformation and, therefore, is not reliant on deformation history. On the other hand, viscoplastic strain is not fully recoverable, resulting in permanent deformation. Therefore, viscoplastic strain does rely on deformation history. Elastic strain is required to determine the internal force vector and is given by Equation 2.5.

$$\varepsilon_e = \varepsilon_{\text{total}} - \varepsilon_{vp(i)} \tag{2.5}$$

Here, $\varepsilon_{vp(i)}$ is the viscoplastic strain at time increment i. The viscoplastic strain is determined through the evaluation of a chosen plasticity model and resolved in a later step of the explicit analysis. Having found the elastic strain, the stress can now be determined through the application of Hooke's law; see Equation 2.6.

$$\sigma = E\varepsilon_e = E\left(\varepsilon_{\text{total}} - \varepsilon_{vp}\right) \tag{2.6}$$

Here E is the stress–strain relationship matrix, or elasticity matrix, which relates stress and strain through the elastic modulus, shear modulus, and Poisson's ratio. Figure 2.1 depicts a typical stress–strain curve. As seen in this figure, the stress is only reliant on the elastic strain and elastic modulus.

Once the stress is calculated, the internal force vector can be found using Equation 2.7, which determines the acceleration in Equation 2.3 through a nodal force balance.

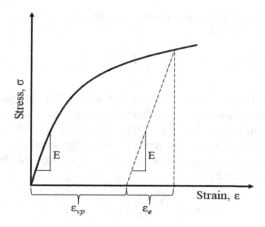

FIGURE 2.1 Typical uniaxial stress–strain curve.

$$P = \int_V B^T \sigma \, dV \tag{2.7}$$

In this equation, V represents the element volume at the current time step.

The viscoplastic strain is determined through the analysis of a flow rule; in this case, the Johnson–Cook (JC) flow rule is used. The JC flow rule condition evaluates the relationship between the JC flow stress and static stress, and it is given by Equation 2.8.

$$\bar{\sigma}_y \geq \bar{\sigma}_0 \tag{2.8}$$

$\bar{\sigma}$ is the JC flow stress, given by Equation 2.9, and $\bar{\sigma}_0$ is the static stress, given by Equation 2.10 [44].

$$\bar{\sigma}_y = \left[A + B\left(\bar{\varepsilon}_{vp}\right)^n \right] \left[1 + C \ln\left(\frac{\dot{\bar{\varepsilon}}_{vp}}{\dot{\bar{\varepsilon}}_0} \right) \right] \left[1 - \left(T^*\right)^m \right] \tag{2.9}$$

$$\bar{\sigma}_0 = \left[A + B\left(\bar{\varepsilon}_{vp}\right)^n \right] \left[1 - \left(T^*\right)^m \right] \tag{2.10}$$

In the JC equations, A is the initial yield stress, B is the hardening modulus, C is the strain rate-dependent coefficient, m is the thermal softening variable, and n is the work hardening variable. A, B, C, m, and n are material-specific variables determined through experimentation. The first bracketed term accounts for strain hardening or plastic strain accumulation. The second bracketed term accounts for the effects of strain rate. The third, and final, bracketed term accounts for the effects of temperature. $\dot{\bar{\varepsilon}}_0$ is a reference strain rate, which is usually set to 1 or the strain rate for which A and B were determined, and T^* is the homologous temperature, a ratio of the difference between the current temperature and room temperature over the difference between the material melting temperature and the current temperature, Equation 2.11.

$$T^* = \frac{T - T_0}{T_m - T} \qquad (2.11)$$

The JC equations shown here are represented as one-dimensional functions when they are, in fact, multi-dimensional but hold the same relationships as presented. The flow stress equation, Equation 2.9, is a function of the von Mises stress equation. Stress can be divided into two categories: hydrostatic stress and deviatoric stress. Hydrostatic stress is the pressure stress and is the average of the three principal stresses; see Equation 2.12 [45].

$$\sigma_{\text{hyd}} = \frac{1}{3}(\sigma_{11} + \sigma_{22} + \sigma_{33}) \qquad (2.12)$$

Based on the assumption that metals are incompressible, it follows that hydrostatic stress cannot cause deformation. Meaning that deformation is caused by the deviatoric stress, which is the total stress minus the hydrostatic stress presented in Equation 2.13 [45].

$$\sigma' = \sigma_{\text{tot}} - \sigma_{\text{hyd}} \qquad (2.13)$$

The von Mises stress is derived from the deviatoric stress, as shown in Equation 2.14.

$$\sigma_{vM} = \left(\frac{3}{2}\sigma' : \sigma'\right)^{1/2} \qquad (2.14)$$

where : is the double inner product, carried out across two second-order stress tensors, which provides a scalar output. This formulation is useful because it provides an invariant effective stress that can be used to predict the onset of yield, which is why the von Mises stress is also referred to as the effective stress, σ_e [39].

If the JC flow rule Condition evaluates as true, then yielding has occurred, and the viscoplastic strain rate must be determined to calculate the equivalent viscoplastic strain. Otherwise, yielding has not occurred, and the element is still in the elastic region, and, therefore, the viscoplastic strain rate is zero. The viscoplastic strain rate is determined through Equation 2.15.

$$\dot{\bar{\varepsilon}}_{vp(i+1/2)} = \dot{\varepsilon}_0 \exp\left[\frac{1}{C}\left(\frac{\bar{\sigma}_y}{\bar{\sigma}_0} - 1\right)\right] \qquad (2.15)$$

The viscoplastic strain rate is then used to calculate an equivalent plastic strain; see Equation 2.16.

$$\bar{\varepsilon}_{vp(i+1)} = \bar{\varepsilon}_{vp(i)} + \sqrt{\frac{2}{3}\dot{\bar{\varepsilon}}_{vp(i+1/2)} : \dot{\bar{\varepsilon}}_{vp(i+1/2)}} \, \Delta t_{(i+1/2)} \qquad (2.16)$$

Here is the double inner product again, and $\Delta t_{(i+1/2)}$ is the half time-step increment. This equivalent plastic strain is carried forward to the next time step and used to determine the elastic strain. This time-step iteration process is continued for all the subject finite elements until the time interval is exhausted or the elements are damaged, which is covered in Section 2.2 [43]. The explicit analysis CDTI process is represented by the flow chart depicted in Figure 2.2.

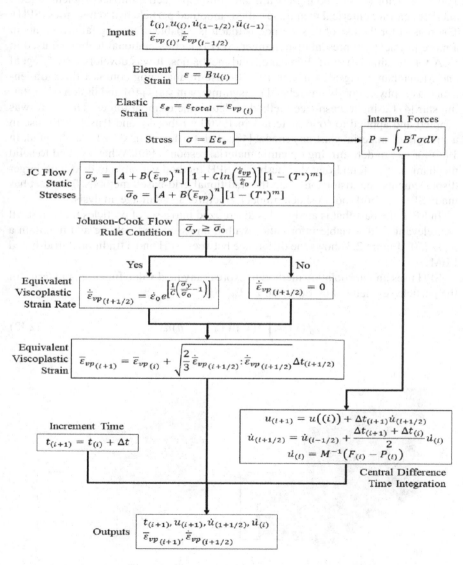

FIGURE 2.2 Explicit analysis central difference time integration flow chart.

An equation of state (EOS) is required when modeling dynamic impacts to balance the projectile's physical properties and target due to shock wave creation and propagation. For this research, the Mie-Grüneisen EOS will be used. A further discussion on impact shock waves and EOSs can be found in Section 2.1.3.

2.1.2 Smoothed Particle Hydrodynamics

The large deformation and high strain rates of the projectile impact problem require another central numerical technique called smoothed particle hydrodynamics (SPH). The reason for the use of this method is that it is capable of tracing failure events in a more physically representative manner. SPH is a computational technique used in FEA for the simulation of fluids and solid mechanics. It was developed by Gingold and Monaghan alongside Lucy in 1977 for use in analyzing complex three-dimensional astrophysics problems related to asymmetry in stars [46] and fission of a rotating star [47] using a mesh-free method to determine element forces. This work was adapted and applied to solid materials in 1990 by Libersky and Patschek for use in a strength of material elasticity model [48], then further refined by Libersky et al. in 1993 for use in determining dynamic material response [49]. When applied to solid mechanics, they found that the mesh-free SPH method was able to deal with larger displacements and distortions than the traditional grid-based methods. This fact has made SPH a useful tool in shock, impact, fracture, and damage analysis.

In SPH, the part that is analyzed is discretized into a set of particles that retain all the relevant field variable information within an associated volume and maintain a mass [50]. Figure 2.3 shows the difference between SPH and a traditional grid-based FEM.

SPH uses an interpolation method to express any field value function, f in terms of the particle set, defined by Equation 2.17.

$$f(x) = \int f(x')W(x - x',h)\,dx' \qquad (2.17)$$

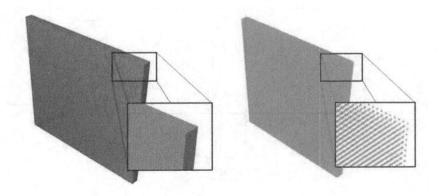

FIGURE 2.3   Grid versus smoothed particle hydrodynamics mesh.

where x is the point of interest, x' denotes the particle positions, h is the smoothing length variable, and W is the interpolating kernel, which must satisfy the two properties presented in Equations 2.18 and 2.19.

$$\int W(x - x', h)\, dx' = 1 \tag{2.18}$$

$$\lim_{h \to \infty} W(x - x', h) = \delta(x - x') \tag{2.19}$$

where δ is the delta function [51]. In the original work by Gingold and Monaghan, calculations were performed using a Gaussian kernel, shown in Equation 2.20 [46].

$$W(x - x', h) = \frac{1}{h\sqrt{\pi}}\, e^{-\frac{(x-x')^2}{h^2}} \tag{2.20}$$

which provides the benefits of being C^∞ continuous, but with the Gaussian kernel, all particles within the SPH mesh contribute to the calculation. What has become more of a standard use kernel is the cubic spline kernel. The cubic spline is the kernel used within the Abaqus software package, which was the primary FEA software used in this research. The cubic spline kernel is shown in Equation 2.21 [43].

$$W(x - x', h) = \frac{1}{h^3 \pi} \begin{cases} 1 - \dfrac{3}{2}\xi^2 + \dfrac{3}{4}\xi^3, & \text{for } 0 \le \xi \le 1 \\[2mm] \dfrac{1}{4}(2 - \xi)^3, & \text{for } 1 \le \xi \le 2 \\[2mm] 0, & \text{for } \xi > 2 \end{cases} \tag{2.21}$$

where $\xi = \dfrac{x}{h}$.

The cubic spline kernel reduces the number of particles included in the calculations to those within twice the smoothing length of the particle of interest while still maintaining C^2 continuity. Regardless of the kernel chosen, for numerical operations, the field variables can be approximated through a kernel summation, which is presented in Equation 2.22.

$$f(x) \cong \sum_{k=1}^{N} f_k W(x - x_k, h) \frac{m_k}{\rho_k} \tag{2.22}$$

where k represents the particle index, N is the total number of particles, f_k is the field variable at the kth particle, x_k are the particle's position, W and k remain the kernel function and smoothing length variable, m_k is the mass of the kth particle, and ρ_k is the density of the kth particle [52]. A visual example of a kernel function is depicted in Figure 2.4, where the solid particles would be included in the computations, and empty particles would not be included.

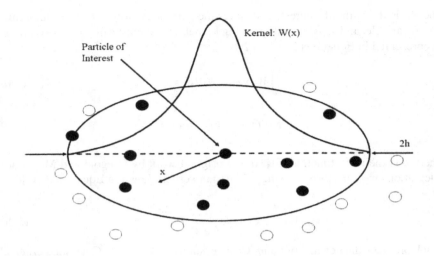

FIGURE 2.4 Smoothed particle hydrodynamics kernel function.

As shown in Equations 2.17 and 2.22, one of the primary differences between SPH and traditional grid-based FEM is that SPH solves the field value differential functions by discretizing the part volume into particles versus using a point-wise discretization of space-time [53]. Because the particles contain volume and mass properties, they cannot be analyzed the same way as a traditional grid mesh under impact conditions. Instead of solely using the conventional explicit technique outlined in Section 2.1.1, the three conservation equations are used to determine particle response [51]. SPH still uses a Lagrangian formulation; however, in the SPH implementation, the conservation equations must be satisfied for each particle at each time increment. The updated field values will be carried forward to the next time step.

The first conservation equation that must be addressed is the conservation of mass, or continuity equation, due to the density determined through the continuity equation being used to evaluate the remaining two conservation equations. In evaluating the interaction between a particular particle pair, notated as $a - b$, the conservation of mass equation can be represented as shown in Equation 2.23, and the time rate of change of mass density is shown in Equation 2.24. Subscript a denotes the properties at the particle of interest, and subscript b represents the properties at another particle within the field.

$$p_a = \sum_b m_b w_{ab} \tag{2.23}$$

$$\frac{dp_a}{dt} = \sum_b m_b (v_a - v_b) \nabla_a W_{ab} \tag{2.24}$$

where ρ is the particle density, m is the individual mass of the particle, v is the velocity of the particle, ∇ is the gradient function taken with respect to the coordinates of the particle of evaluation, and W_{ab} is the kernel function relating particles a and b.

Once the conservation of mass equation has been satisfied, the conservation of momentum, or momentum equation, will be evaluated. Again, in assessing the interaction between a particle pair $a - b$, the momentum equation starts from a pressure gradient estimate, shown in Equation 2.25.

$$p_a \nabla P_a = \sum_b m_b (P_b - P_a) \nabla a W_{ab} \tag{2.25}$$

Here, P is the pressure at the particle. To symmetrize the pressure gradient, the equation can be rewritten using the relationship found in Euler's equation, Equation 2.26, to arrive at Equation 2.27 [51].

$$\frac{dv}{dt} = -\frac{\nabla P}{\rho} = -\nabla \left(\frac{P}{\rho} \right) + \frac{P}{\rho} \nabla \rho \tag{2.26}$$

$$\frac{dv_a}{dt} = -\sum_b m_b \left(\frac{P_b}{p_b^2} + \frac{P_a}{p_a^2} \right) \nabla_a W_{ab} \tag{2.27}$$

where dv/dt is the time derivative of the velocity vector. By using this form of the momentum equation, both the linear and angular momenta are conserved, which may not be the case with an asymmetric pressure gradient term.

Finally, the conservation of energy, or energy equation, can be addressed. In the SPH formulation presented by Gingold and Monaghan, this is the rate of change of thermal energy per unit mass of the particles, or the hydrodynamic energy equation, shown in Equation 2.28 [46].

$$\frac{de}{dt} = -\left(\frac{P}{p} \right) \nabla \cdot \vec{v} \tag{2.28}$$

where de/dt is the time derivative of specific internal energy, e. This equation can be rewritten to determine the conservation of energy at particle a and simplified for adiabatic systems, as shown in Equation 2.29.

$$\frac{d\varepsilon_a}{dt} = \left(\frac{P_a}{p_a^2} \right) \sum_b m_b (\vec{v}_a - \vec{v}_b) \cdot \nabla_a W_{ab} \tag{2.29}$$

Several variations on the energy equation can be used in SPH analysis, and it is important to note that several of these forms can present non-realistic physical solutions, such as negative internal energy. These issues are typically solved through the use of a predictor–corrector approach for analysis. In this approach, the governing conservation equations predict the field variables using the chosen kernel, the predictor phase. This prediction may lead to an unbalanced energy solution, which is corrected through a local restoration of the conservation of energy equation, the corrector phase. Following the corrector phase, the field values are adjusted to meet the new particle state. This process is followed at every time step used in the analysis.

With a sufficiently small time increment, the numerical adjustment used in the predictor–corrector method does not affect the accuracy of the solution [52].

One of the concerns that arose from adapting SPH to solid mechanics is the issue of boundary effects. When the summation approximation, Equation 2.28, is applied near a boundary, there is a truncation error, which results in an incomplete summation, and C^0 continuity may not be maintained. This error means that rigid body motion may not be determined correctly through the analysis process [53]. There have been many efforts to overcome this error to regain at a minimum C^1 continuity through the use of Lagrangian stabilization [54], symmetric formulation [55,56], Galerkin formulation [57], least squares methods [58,59], or ghost particle method [60].

Abaqus incorporates the ghost particle method to deal with SPH boundary surfaces. This method creates imaginary particles when an SPH body interacts with a solid Lagrangian boundary. Interaction is considered when a particle is within twice the smoothing length of a boundary surface. In this case, a virtual plane is formed along this boundary, with the ghost particles being formed across the plane from the physical particles. The ghost particles' field properties are computed from those of the physical particles as if the SPH part spanned the virtual plane, but they are assigned the opposite sign of the physical particles. The opposing field values produce a Leonard–Jones molecular force, which forms a repulsion force along the boundary surface, preventing SPH particle penetration of the solid [61]. The use of ghost particles improves the performance of SPH method integration along the boundaries, and it is worth noting that the ghost particles are not permitted to interact back across the virtual plane with the physical particles. The ghost particles are still permitted to interact among themselves, and as noted by Colagrossi and Landrini, care must be taken to avoid an excess of ghost particle mass [60]. An excess of ghost particle mass can start providing false inputs into the boundary surface integration, which in turn can lead to errors in physical particle field values along the boundary. To limit the ghost particle mass, the smoothing length can be decreased near a boundary, which is the application method used within Abaqus. See Figure 2.5 for a pictorial representation of the ghost particle method acting along a perpendicular boundary surface.

Figure 2.6 depicts the explicit SPH process incorporating the three conservation equations plus an equation of state, which will be discussed in Section 2.1.3.2.

2.1.3 Shock Waves and Equation of State

2.1.3.1 Shock Waves
A compressive shock wave is generated under the loading of high strain rate impact, which propagates through both the target and projectile. Across this shock wave, a discontinuity in material properties is developed, which must be accounted for through the conservation equations. The Rankine–Huguenot equations for fluids can be adapted for solid materials, providing simplified equations relating the conditions immediately before and after the wave. Figure 2.7 presents a simple one-dimensional schematic of a wave propagating through a material as a baseline case for the simplified relations developed by Meyers in *Dynamic Behavior of Materials* [62].

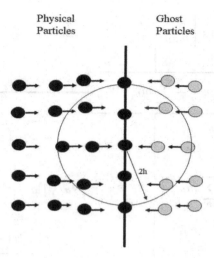

FIGURE 2.5 Smoothed particle hydrodynamics ghost particles.

In this figure, P_i is the particle pressure, v_i is the particle velocity, ρ_i is the particle density, and e_i is the particle internal energy. The subscript 0 indicates the values still within the elastic regime, ahead of the wave, and the subscript 1 indicates the values within the plastic regime, behind the wave. v_w represents the wave velocity. The conservation of mass equation regarding the balance across the shock wave becomes a mass flow balance through the shock wave, where the mass in must equal the mass out. The mass going into the wave front can be represented by Equation 2.30, and the mass going out of the wave can be represented by Equation 2.31.

$$A\rho_0 (v_w - v_0)dt \tag{2.30}$$

$$A\rho_0 (v_w - v_1)dt \tag{2.31}$$

This representation provides the conservation of mass equation seen in Equation 2.32 [62].

$$A\rho_0 (v_w - v_0) = A\rho_1 (v_w - v_1) \tag{2.32}$$

Because the cross-sectional area, A, and wave time step, dt, do not change during shock wave propagation, Equation 2.32 can be simplified to Equation 2.33.

$$\rho_0 (v_w - v_0) = \rho_1 (v_w - v_1) \tag{2.33}$$

Since the particle velocity before and after the wave is not the same, the shock wave propagation through a material leads to a change in particle density. For the case of

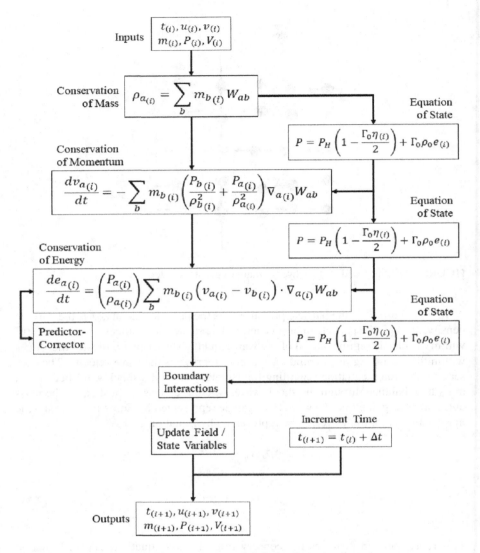

FIGURE 2.6 Explicit analysis smoothed particle hydrodynamics flow chart.

the target, where $v_0 = 0$, the conservation of mass equation can be further simplified to Equation 2.34.

$$\rho_0 v_w = \rho_1 (v_w - v_1) \tag{2.34}$$

The conservation of momentum accounts for the momentum flow across the wave, where the difference in momentum, Equation 2.35, must equal the wave impulse, Equation 2.36 [62].

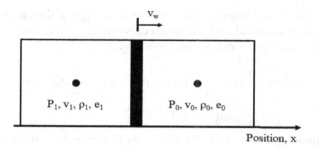

FIGURE 2.7   Wave propagation through a material.

$$\Delta M = \rho_1 A (v_w - v_1) dt \, v_1 - \rho_0 A (v_w - v_0) dt \, v_0 \tag{2.35}$$

$$\text{Im} \, p = F \, dt = (P_1 - P_0) A \, dt \tag{2.36}$$

Again, the cross-sectional area and wave time step are constant, so by equating the wave impulse with the change in momentum, Equation 2.37 is obtained [62].

$$P_1 - P_0 = \rho_0 (v_w - v_0)(v_1 - v_0) \tag{2.37}$$

This form of the momentum equation indicates that the change in pressure across a wave boundary is equal to the change in momentum across the same wave. Again, accounting for the case of the target, $v_0 = 0$, the conservation of momentum equation reduces to Equation 2.38.

$$P_1 - P_0 = \rho_0 v_w v_p \tag{2.38}$$

Here, the term $\rho_0 v_w$ is often referred to as the shock impedance of the material. The conservation of energy equation creates a balance between the change in external work done across the wave, Equation 2.39, and the change in total energy across the wave, Equation 2.40.

$$\Delta W = (P_1 A)(v_1 dt) - (P_0 A)(v_0 dt) \tag{2.39}$$

The first term, $(P_i A)$, equates to the force, and the second term, $(v_i A)$ represents the distance.

$$\Delta E_{\text{total}} = \frac{1}{2} \Big[\rho_1 A (v_w - v_1) dt \Big] v_1^2 + e_1 A \rho_1 (v_w - v_1) dt -$$

$$\left\{ \frac{1}{2} \Big[\rho_0 A (v_w - v_0) dt \Big] v_0^2 + e_0 A \rho_0 (v_w - v_0) dt \right\} \tag{2.40}$$

In Equation 2.40, e_i indicates the specific internal energy of the particle. By equating ΔW and ΔE, and considering a constant cross-sectional area, the conservation of energy equation can be represented by Equation 2.41.

$$P_1 v_1 - P_0 v_0 = \frac{1}{2}\rho_1(v_w - v_1)v_1^2 + e_1\rho_1(v_w - v_1) - \frac{1}{2}\rho_0(v_w - v_0)v_0^2$$
$$- e_0\rho_0(v_w - v_0) \tag{2.41}$$

Using the conservation of mass relationship in Equation 2.33, the conservation of energy equation can be reduced to Equation 2.42.

$$P_1 v_1 - P_0 v_0 = \frac{1}{2}\rho_0(v_w - v_0)(v_1^2 - v_0^2) + \rho_0(v_w - v_0)(e_1 - e_0) \tag{2.42}$$

For the target, $V_0 = 0$, the conservation of energy equation further simplifies to Equation 2.43.

$$P_1 v_1 = \frac{1}{2}\rho_0 v_w v_1^2 + \rho_0 v_w(e_1 - e_0) \tag{2.43}$$

The three conservation equations are not sufficient to solve for the four variables that represent the material's state following a compressive shock wave. In this case, an EOS is used to relate these variables.

2.1.3.2 Equation of State

An EOS provides a hydrodynamic material model for use in FEA, in which the system's pressure, volume or density, temperature, and energy are related through the application of a specific EOS [63]. EOS is used to relate the state variables together when the conservation equations are not enough, which is the case of high strain rate impact. One of the most commonly used EOS in impact modeling is the Mie-Grüneisen equation, which has also proven effective at predicting the response of porous materials to compressive shock waves [62,64–66]. To develop a relationship between pressure and volume, there must be a link between statistical mechanics and thermodynamics, which starts with the analysis of the Helmholtz free energy equation, Equation 2.44 [62].

$$A_H = U - TS \tag{2.44}$$

where A_H is the Helmholtz free energy, U is the internal energy of the system, T is the absolute temperature of the surroundings, and S is the entropy of the system. The Helmholtz free energy is a measure of the work attainable within a closed thermodynamic system with constant temperature, isothermal. Using the fundamental thermodynamic relationship, see Equation 2.45, pressure is obtained by taking the differential of the Helmholtz free energy with regard to volume, Equation 2.46 [67].

$$dA = T\,dS - F\,dV \tag{2.45}$$

$$P = \left| \frac{\partial A_H}{\partial V} \right|_T \tag{2.46}$$

Through the use of the virial theorem, mathematically represented in Equation 2.47 [68], Grüneisen obtained the pressure–energy relationship shown in Equation 2.48 [69]. Clausius first defined the virial theorem in 1870 through his efforts to analyze the local stresses around a single atom. Virial comes from the Latin "vis," which means energy or force [70]. For mechanics, the virial theorem states that twice the total kinetic energy within a system is equal to the virial of the forces within the total system [68].

$$KE = -\frac{1}{2} \sum_{k=1}^{N} (F_k \cdot x_k) \tag{2.47}$$

where KE represents the total kinetic energy, N is the total number of particles, F_k is the force applied to the kth particle, and x_k is the position of particle k.

$$P = \frac{\Gamma}{V} E \tag{2.48}$$

where Γ is the Grüneisen coefficient, the ratio of the vibration frequency, v, to volume V at a constant temperature, Equation 2.49 [71],

$$\Gamma = \frac{d \ln(v)}{d \ln(V)} \tag{2.49}$$

From here, Mie simplified the equation by relating the current conditions to those of a known value, either at 0 K or from a point of the Huguenot curve. A further simplification was made relating the volume to the density of the material. This simplification led to the Mie-Grüneisen equation taking a linear form with respect to energy, as shown in Equation 2.50 [43].

$$P - P_H = \Gamma \rho (e - e_H) \tag{2.50}$$

Here P is the current pressure, P_H is the Huguenot pressure, Γ is the redefined Grüneisen ratio, ρ is the material density, e is the specific internal energy, and e_H is the Huguenot specific energy. The Grüneisen ratio is presented in Equation 2.51, the Huguenot pressure is shown in Equation 2.52, and the Huguenot specific energy is shown by Equation 2.53.

$$\Gamma = \Gamma_0 \frac{\rho_0}{\rho} \tag{2.51}$$

where Γ_0 is the material property, r_0 is the reference density, and ρ is the current density.

$$P_H = \frac{\rho_0 r_0^2 \eta}{\left(1 - s_H \eta\right)^2} \qquad (2.52)$$

where c_0 is the material reference speed of sound, η is a nominal compressive strain, and s_H is the linear Huguenot slope coefficient.

$$e_H = P_H \frac{\eta}{2\rho_0} \qquad (2.53)$$

In Equations 2.52 and 2.53, the nominal strain is represented by Equation 2.54.

$$\eta = 1 - \frac{\rho_0}{\rho} \qquad (2.54)$$

Abaqus uses an alternative form of the Mie-Grüneisen equation in its analysis, as shown in Equation 2.55 [43].

$$P = P_H \left(1 - \frac{\Gamma_0 \eta}{2}\right) + \Gamma_0 \rho_0 e \qquad (2.55)$$

This equation is solved simultaneously alongside the three conservation equations at each material point and time increment to ensure that the state variables are balanced throughout an impact event. In the implementation of the Mie-Grüneisen EOS, the reference material properties used are those for Inconel 718 (IN718), as that is the base material of the designs, and the current density will be the relative density of the periodic cellular design.

2.2 DAMAGE MODELING

As mentioned in Section 2.1.1, a damage model is required to determine when the elements or particles fail within a simulation to be analyzed appropriately as part of the FEA process. It is important to point out that the exploration in uses of these advanced materials under high strain rate applications is in its infancy. Not a great deal has been accomplished in this field, but the association with its future development will allow for an appreciation of what can be developed with future research. Two different failure models were considered as part of this research. The first model, the JC failure model, is predominantly used to model the failure of metals. The second model, the Holmquist–Johnson–Cook (HJC) failure model, is used to model failure in brittle materials, such as concrete. Each of these models will be described in more detail.

2.2.1 JOHNSON–COOK FAILURE MODEL

The JC failure model was developed in the 1980s and fully described in 1985 [72]. It makes the assumption that the change in material properties between static and dynamic cases is due to strain rate effects, which can account for large strains, high temperatures, and high pressures. It incorporates the JC material model and is based on plastic damage accumulation. This model is commonly used for estimating the dynamic deformation of metals under high strain rates [73–75].

The JC failure model defines material damage as the sum of incremental equivalent plastic strain divided by the critical fracture strain; see Equation 2.56.

$$D = \sum \frac{\Delta \overline{\varepsilon}_f}{\overline{\varepsilon}_{cr}} \tag{2.56}$$

In this equation, D is the JC damage coefficient, $\overline{\varepsilon}_f$ is the JC fracture strain, and $\overline{\varepsilon}_{cr}$ is the critical fracture strain. The right-hand side of Equation 2.56 sums the incremental change in the element or particle plastic strain, see Equation 2.57, and compares it as a ratio to the critical failure strain of the material, which is presented as the JC damage coefficient. The damage coefficient has a range of zero to one, where zero represents a pristine, or undamaged material, and one represents the material being fully damaged, and fracture will occur in metals.

$$\overline{\varepsilon}_f = \left[D_1 + D_2 e^{D_3 \sigma^*} \right] \left[1 + D_4 \varepsilon^* \right] \left[1 + D_5 T^* \right] \tag{2.57}$$

Here, the first bracketed term contains the stress triaxiality effects. D_1, D_2, and D_3 are material-specific model-fitting properties, and σ^* is the ratio of the average normal stress to von Mises equivalent stress. The second bracketed term comprises the strain rate effects. D_4 is another material-specific model-fitting property, and ε^* is the dimensionless strain rate ratio of viscoplastic strain rate to reference strain rate. The reference strain rate here is the same as that used in Equation 2.9. The final bracketed term includes the effects of temperature on material failure. D_5 is a material-specific model-fitting parameter, and T^* is the material's homologous temperature, Equation 2.11. The D_i terms are traditionally found through experimentation, but in this case, a combination of experimentation and computational optimization will be used, which is described in more detail in Chapter 3. Figure 2.8 depicts the JC damage model implementation as a part of an explicit analysis technique. Using the CDTI analysis technique depicted in Figure 2.2, the JC damage analysis shown here would branch off from the JC flow stress, and static stress step, shown in both figures, whereas using the SPH analysis technique depicted in Figure 2.6, the inputs would come from the field and state variable updates before beginning the next time step.

2.2.2 HOLMQUIST–JOHNSON–COOK FAILURE MODEL

Although the JC model works well for metals, such as the IN718 that will be used to manufacture the projectile, it does not model the brittle failure modes of materials such as concrete well, of which the target will be modeled. The HJC failure model was developed in the late 1980s and fully described in 1993 [76]. It makes the assumption that the normalized strength of the material can be determined through the element pressure and strain rate. It incorporates the HJC strength model, which is presented in Equation 2.58. This model is commonly used to estimate the dynamic deformation of brittle materials subject to large strains, high strain rates, and high pressures [77,78].

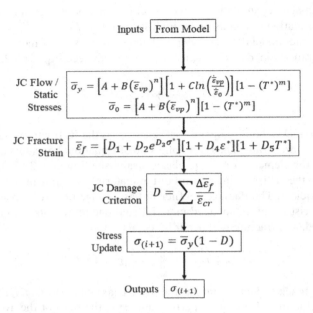

FIGURE 2.8 Johnson–Cook failure model flow chart.

$$\sigma^* = \left[A(1 - D) + B\,P^{*^n} \right]\left[1 + C\ln\left(\dot{\varepsilon}^*\right) \right] \tag{2.58}$$

In this representation, σ^* is the normalized strength of the material and defines as the ratio of the von Mises equivalent stress and uniaxial compressive strength. A is the normalized pressure, B is the normalized pressure coefficient, C is the strain rate coefficient, and n is the pressure hardening exponent; all these variables are experimentally derived material constants. P^* is the normalized pressure ratio of element pressure to axial compressive strength, and $\dot{\varepsilon}^*$ is the dimensionless strain rate ratio of equivalent viscoplastic strain rate to the reference strain rate. Finally, D is the HJC damage coefficient, which is described next. The HJC model defines metal damage as the sum of incremental equivalent elastic strain and volumetric strain divided by the critical fracture strain; see Equation 2.59.

$$D = \sum \frac{\Delta\bar{\varepsilon}_{vp} + \Delta\mu_{vp}}{\bar{\varepsilon}_f} \tag{2.59}$$

D is the HJC damage coefficient mentioned above. ε_{vp} is the equivalent viscoplastic strain, μ_{vp} is the volumetric compaction strain, and $\bar{\varepsilon}_f$ is the HJC fracture strain. The right-hand side of Equation 2.59 sums the incremental change in the element or particle viscoplastic strain and compares it as a ratio to the HJC fracture strain of the material, which is presented in Equation 2.60. Again, the damage coefficient has a range of zero to one. Again, zero represents a pristine or undamaged material, and one represents the material being fully damaged, and fracture of the material occurs.

$$\overline{\varepsilon}_f = D_1 \left(P^* + T^* \right)^{D_2} \tag{2.60}$$

In the HJC fracture strain equation, D_1 and D_2 are material-specific model-fitting properties, P^* is the same normalized pressure ratio as in the HJC strength model, and T^* is the normalized tensile hydrostatic strength ratio of the tensile strength to the uniaxial tensile strength. As with the JC failure mode, the D_i terms are found through experimentation. Since concrete is not a focus of this research, the failure parameters for concrete will be taken from previously established studies [76,79].

2.3 CONSTITUTIVE MODELS

This research will incorporate the use of several constitutive models to provide the necessary relations in three different aspects of the material response of the periodic cellular structures. Three of the models have already been introduced. The JC flow stress constitutive model will be used in conjunction with a lattice model within the FEA explicit approach to determine the viscoplastic response of the designs. The JC and HJC damage models will be used to determine the damage and failure of elements or particles within the FEA model. The remaining two constitutive models, which will be described in more detail, are used to determine the static and near-static equivalent material properties of the different structural designs. These models were foundation elements in determining a new dynamic lattice model for use in FEA. The material properties determined through these constitutive relations will serve as inputs into the FEA model for the respective cellular designs.

2.3.1 CONSTITUTIVE COMPRESSION RESPONSE MODELS

To determine equivalent material properties for variations in relative density of a base periodic structure, a constitutive compression response model that accurately describes the structure's compressive response must be found. Once a model is developed, it can be used to determine the properties for variations on the experimentally tested designs for use in computational analysis. This model will account for making changes to the surface thickness, and therefore relative density, to optimize a design for high strain rate impact.

It is worth noting that the point at which metal lattice designs transition from a quasi-static strain rate response to a dynamic strain rate response has yet to be determined. Work accomplished using polymer foams has indicated a higher degree of influence in the mechanical response from strain rates greater than approximately 1,000/sec [80]. To determine the strain rate-dependent transition point, uniaxial compression testing will occur at multiple strain rates from a baseline static rate based on the American Society for Testing and Materials (ASTM) E9-19 standard rate of 0.005 in/in/min rate, adjusted for specimen dimensionality, and increasing the rate to near the usable limit of the MTS Systems Corporation machine. Those data will be used to determine a boundary of where the constitutive compression models fit the data, thereby determining where static and quasi-static conditions hold and where the response is considered dynamic.

The models developed for analyzing the compressive response of cellular materials are divided into two categories, phenomenological models and micro-mechanical models. Phenomenological models are designed to provide the best fit to experimental results of the mechanical behavior without consideration of any physical phenomenon. Micro-mechanical models are developed with a basis in the study of deformation mechanism of the cell structure under loading. The Rusch model is a phenomenological model that will be used, and the Gibson modified model is a micro-mechanical model that will be used for analysis in this research.

2.3.1.1   Rusch Model

As mentioned, the Rusch model is a phenomenological model developed in the late 1960s based on his work with polymer foams [81,82]. Still, it has been used effectively with a variety of materials [83–85]. In this research, it will be applied to the experimentally derived responses of metal lattice structures to develop a model for further analysis. The Rusch model has a relatively simple formulation consisting of a summation of two power laws, see Equation 2.61, but has proven effective in modeling the stress–strain relationship of cellular solids under compression loading [84,85].

$$\sigma = A\varepsilon^m + B\varepsilon^n \qquad (2.61)$$

Here, σ and ε represent the engineering stress and engineering strain, respectively. The first power law term is used to fit the model to the plateau stress region of the response. Here, A is the relative density–dependent and material-specific model-fitting coefficient, and m is the material-specific fitting exponent independent of the design relative density. The second power law term is used to fit the densification region of the material response. Again, B is the material-specific model coefficient dependent on relative density, and n is the material-specific model-fit exponent independent of relative density. m is found within the range of zero to one, while n is found to be greater than one. A and B can be approximated by power laws as well, see Equations 2.62 and 2.63, with the relationship based on empirical data comparing the coefficients across multiple relative densities [84].

$$A = C_{1,A} r\, d^{C_{2,A}} \qquad (2.62)$$

$$B = C_{1,B} r\, d^{C_{2,B}} \qquad (2.63)$$

The benefit of the Rusch model is that it only requires four parameters. Yet, it does exhibit the drawback that it will typically either match the plateau stress region or the densification region, but not both. A representation of the relationship between the plateau stress and densification terms of the Rusch model can be seen in Figure 2.9.

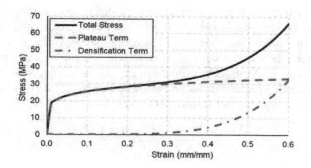

FIGURE 2.9   Rusch model relationship.

2.3.1.2   Gibson Modified Model

The Gibson modified model [83] is an extension of the original Gibson model [6], both being micro-mechanical models based on the deformation theory. However, in the case of this research, the Gibson modified model will be fit to experimentally derived response curves, and model coefficient relationships will be developed specific to a lattice design while allowing for variation in cellular design parameters. The Gibson models define three regions of the compression stress–strain response. These are the linear elastic region, the plateau region, and the densification region. Modification in the plateau region was updated to account for a positive slope in this region. The model is presented for the linear elastic region in Equation 2.64, the modified plateau region in Equation 2.65, and the densification region in Equation 2.66.

$$\sigma = E\varepsilon \tag{2.64}$$

$$\sigma = \sigma_y + h\varepsilon \tag{2.65}$$

$$\sigma = \sigma_y \frac{1}{D}\left(\frac{\varepsilon_D}{\varepsilon_D - \varepsilon}\right)^m \tag{2.66}$$

where in each of the equations σ and ε are the engineering stress and strain, respectively. In Equation 2.64, E is the elastic modulus. For the case of open cell foams and lattices, Gibson and Ashby determined that the elastic modulus is unique for a specific cellular design but is a function of density only. This relationship can be represented by Equation 2.67 [6].

$$E = C_E\rho^2 \tag{2.67}$$

C_E is the coefficient of elasticity, which can be determined through curve-fitting experimental data. Once C_E is known, E values can be approximated for changes in density under the same cellular design. In Equation 2.65 σ_y is the yield stress, and h is the Gibson modified plateau slope coefficient, which is determined through experimentation. Gibson and Ashby determined that the yield stress for open cell structures was again unique to a particular design but still a function of density, and it arrived at the representation shown in Equation 2.68 [6].

$$\sigma_y = C_y \rho^{\frac{3}{2}} \tag{2.68}$$

Here, C_y is the yield coefficient, which is determined through curve-fitting experimental data. Once again, after C_y is known, it can be used to approximate the ε_D material response of the cellular design with different densities. In Equation 2.66 D and m are material-specific model-fit parameters that are density independent and determined through experimentation. ε_D is the densification strain, which is again a function of the cellular design and density and can be represented by Equation 2.68 [6].

$$\varepsilon_D = 1 - C_D \rho \tag{2.69}$$

Here, C_D represents the densification coefficient, which is obtained through experimentation. As mentioned, the densification coefficient is unique to a cellular design, and once known, it can be used to model changes in design density. This coefficient can also be approximated by the ratio of 1.4 over the material density of the structure, ρ_s, shown in Equation 2.70 [84].

$$C_D = \frac{1.4}{\rho_s} \tag{2.70}$$

The benefit of the Gibson modified model is that it segments the material response into three regions found in experimentation: linear elastic region, plateau region, and densification region. This allows the model to be tuned to more closely match

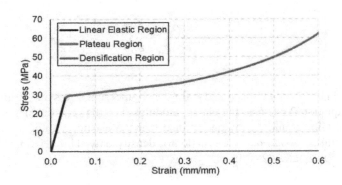

FIGURE 2.10 Gibson modified model relationship.

experimental results. However, this model requires six model-fit parameters, which means that additional data collection is required to determine these parameters. A representation of the relationship between the three regions of the Gibson modified model can be seen in Figure 2.10.

REFERENCES

6. L. J. Gibson and M. F. Ashby. *Cellular Solids: Structure and Properties*. Cambridge University Press, Cambridge, England, 2nd edition, 1997.

39. R. D. Cook, D. S. Malkus, M. E. Plesha, and R. J. Witt. *Concepts and Applications of Finite Element Analysis*. John Wiley & Sons, Inc., Hoboken, New Jersey, 2002.

42. K. J. Bathe. *Finite Element Procedures*. Klaus-Jurgen Bathe, United States, 2nd edition, 2014.

43. Dassault Syst`emes. Abaqus version 6.12 theory manual, 2016.

44. G. R. Johnson and W. H. Cook. A constitutive model and data for metals subjected to large strains, high strain rates, and high temperatures. In *Proceedings of the Seventh International Symposium on Ballistics*, pages 541–547, The Hague, 1983.

45. I. H. Shames and F. A. Cozzarelli. *Elastic and Inelastic Stress Analysis*. Taylor & Francis, Boca Raton, FL, 1997.

46. R. A. Gingold and J. J. Monaghan. Smoothed particle hydrodynamics: Theory and application to non-spherical stars. *Monthly Notices of the Royal Astronomical Society*, 181(3): 375–389, 1977.

47. L. B. Lucy. A numerical approach to the testing of the fission hypothesis. *The Astronomical Journal*, 82(12): 1013, 1977.

48. L. D. Libersky and A. G. Petschek. Smooth particle hydrodynamics with strength of materials. In Harold E. Trease, editors, *Advances in the Free-Lagrange Method Including Contributions on Adaptive Gridding and the Smooth Particle Hydrodynamics Method*, pages 248–257. Springer Berlin Heidelberg, Berlin, Heidelberg, 1990.

49. L. D. Libersky, A. G. Petschek, T. C. Carney, J. R. Hipp, and F. A. Allahdadi. High strain Lagrangian hydrodynamics a three-dimensional SPH code for dynamic material response. *Journal of Computational Physics*, 109(1): 67–75, 1993.

50. J. J. Monaghan. An introduction to SPH. *Computer Physics Communications*, 48(1): 89–96, 1988.

51. J. J. Monaghan. Smoothed particle hydrodynamics. *Annual Review of Astronomy and Astrophysics*, 30(1): 543–574, 1992.

52. M. R. I. Islam, S. Chakraborty, and A. Shaw. On consistency and energy conservation in smoothed particle hydrodynamics. *International Journal for Numerical Methods in Engineering*, 116(9): 601–632, 2018.

53. T. Stranex and S. Wheaton. A new corrective scheme for SPH. *Computer Methods in Applied Mechanics and Engineering*, 200(1–4): 392–402, 2011.

55. R. C. Batra and G. M. Zhang. SSPH basis functions for meshless methods, and comparison of solutions with strong and weak formulations. *Computational Mechanics*, 41(4): 527–545, 2008.

56. G. M. Zhang and R. C. Batra. Symmetric smoothed particle hydrodynamics (ssph) method and its application to elastic problems. *Computational Mechanics*, 43(3): 321–340, 2009.

57. L. Cueto-Felgueroso, I. Colominas, G. Mosqueira, F. Navarrina, and M. Casteleiro. On the Galerkin formulation of the smoothed particle hydrodynamics method. *International Journal for Numerical Methods in Engineering*, 60(9): 1475–1512, 2004.

58. G. A. Dilts. Moving-least-squares-particle hydrodynamics - i. consistency and stability. *International Journal for Numerical Methods in Engineering*, 44(8): 1115–1155, 1999.

59. G. A. Dilts. Moving least-squares particle hydrodynamics ii: Conservation and boundaries. *International Journal for Numerical Methods in Engineering*, 48(10): 1503–1524, 2000.

60. A. Colagrossi and M. Landrini. Numerical simulation of interfacial flows by smoothed particle hydrodynamics. *Journal of Computational Physics*, 191(2): 448–475, 2003.

61. A. Danilewicz and Z. Sikora. Numerical simulation of crater creating process in dynamic replacement method by smooth particle hydrodynamics. *Studia Geotechnica et Mechanica*, 36(3): 3–8, 2015.

62. M. A. Meyers. *Dynamic Behavior of Materials*. Wiley, Hoboken, New Jersey, 1994.

63. W. B. Holzapfel. Equations of state and thermophysical properties of solids under pressure. In A. Katrusiak and P. McMillan, editors, *High-Pressure Crystallography*, pages 217–236. Springer Netherlands, Dordrecht, 2004.

64. S. K. Roy, M. Trabia, B. O'Toole, R. Hixson, S. Becker, M. Pena, R. Jennings, D. Somasoundaram, M. Matthes, E. Daykin, and E. Machorro. Study of hypervelocity projectile impact on thick metal plates. *Shock and Vibration*, 2016: 1–11, 2016.

65. M. A. Zocher, P. J. Maudlin, S. R. Chen, and E. C. Flower-Maudlin. An evaluation of several hardening models using Taylor cylinder impact data. In *European Congress on Computational Methods in Applied Sciences and Engineering (EC-COMAS 2000)*, volume 53, pages 1–20, Barcelona, Spain, 2000. Los Alamos National Laboratory, NM.

66. M. H. Rice, R. G. McQueen, and J. M. Walsh. Compression of solids by strong shock waves. *Solid State Physics - Advances in Research and Applications*, 6(C): 1–63, 1958.

67. F. Mandl. *Statistical Physics*. Wiley, Hoboken, New Jersey, 2nd edition, 1991.

68. G. Marc and W. G. McMillan. The virial theorem. In R.F. Heuff, J.L. Swift, and D.T. Cramb, editors, *Advances in Chemical Physics*, volume 63, pages 209–361. John Wiley & Sons, Inc., Hoboken, New Jersey, 2007.

69. O. Heuze. General form of the Mie-Gru"neisen equation of state. *Comptes Rendus - Mecanique*, 340(10): 679–687, 2012.

71. E. Gruneisen. Theorie des festen zustandes einatomiger elemente. *Annalen der Physik*, 344(12): 257–306, 1912.

73. R. S. Cowan and W. O. Winer. Frictional heating calculations. *Friction, Lubrication, and Wear Technology: ASM Handbook*, 18: 39–44, 1992.

74. H. W. Meyer and D. S. Kleponis. Modeling the high strain rate behavior of titanium undergoing ballistic impact and penetration. *International Journal of Impact Engineering*, 26(1–10): 509–521, 2001.

75. X. Wang and J. Shi. Validation of Johnson-Cook plasticity and damage model using impact experiment. *International Journal of Impact Engineering*, 60: 67–75, 2013.

76. T. J. Holmquist, G. R. Johnson, and W. H. Cook. A computational constitutive model for concrete subjected to large strains, high strain rates, and high pressures. In *Proceedings of the 14th International Symposium on Ballistics*, pages 591–600, Quebec City, 1993.

77. Y. S. Tai and C. C. Tang. Numerical simulation: The dynamic behavior of reinforced concrete plates under normal impact. *Theoretical and Applied Fracture Mechanics*, 45(2): 117–127, 2006.

78. M. Mohamed, E. Eltehawy, I. Kamal, and A. Aggour. Numerical simulation of projectile penetration in reinforced concrete panels. In *International Conference on Aerospace Sciences and Aviation Technology*, 13(AEROSPACE SCIENCES), Cairo, Egypt, pages 1–12, 2009.

79. G. M. Ren, H. Wu, Q. Fang, and X. Z. Kong. Parameters of Holmquist–Johnson–Cook model for high-strength concrete-like materials under projectile impact. *International Journal of Protective Structures*, 8(3): 352–367, 2017.

80. S. Ouellet, D. Cronin, and M. Worswick. Compressive response of polymeric foams under quasi-static, medium and high strain rate conditions. *Polymer Testing*, 25(6): 731–743, 2006.

81. K. C. Rusch. Load–compression behavior of flexible foams. *Journal of Applied Polymer Science*, 13(11): 2297–2311, 1969.
82. K. C. Rusch. Energy-absorbing characteristics of foamed polymers. *Journal of Applied Polymer Science*, 14(6): 1433–1447, 1970.
83. M. Avalle, G. Belingardi, and R. Montanini. Characterization of polymeric structural foams under compressive impact loading by means of energy-absorption diagram. *International Journal of Impact Engineering*, 25(5): 455–472, 2001.
84. M. Avalle, G. Belingardi, and A. Ibba. Mechanical models of cellular solids: Parameters identification from experimental tests. *International Journal of Impact Engineering*, 34(1): 3–27, 2007.
85. M. Avalle and G. Belingardi. A mechanical model of cellular solids for energy absorption. *Advanced Engineering Materials*, 21(4): 1–7, 2019.

3 Experimental Methodology

This chapter introduces the need for the data required for this research in Section 3.1 and explains the experimental and computational methods that are used to acquire and analyze those data.

3.1 INTRODUCTION

With the end state of performing computational analysis and comparison of different period cellular designs under high strain rate impact loading, a number of parameters and characteristics must be defined. These parameters are presented here, along with the method used to obtain them.

- l – cell edge length. Set during fabrication, verified through measurement.
- t – cell all thickness. Set during fabrication, verified through measurement.
- ρ^* – density. Obtained through measurement of mass, through use of a digital scale, divided by the bounding volume of the specimen.
- ρ_s – density of solid material. Obtained through the fabrication of solid part, then measurement of mass divided by the volume of the specimen.
- rd – relative density. Ratio of ρ^* to ρ_s. (Equation 1.1).
- E – elastic modulus. Obtained through analysis of the stress–strain response under compression loading, the slope of the linear elastic region of the response is calculated using MATLAB.
- σ_y – yield strength. Obtain through analysis of stress–strain response under compression loading, using the 0.2% offset method. This property is calculated using MATLAB or Microsoft Excel.
- ε_d – densification strain. Obtained through analysis of the stress–strain response under compression loading, determined as the point where the slope of the densification region of the response curve crosses the abscissa. Densification strain is depicted in Figure 3.5. This property is calculated using MATLAB or Microsoft Excel.
- W – toughness. Obtained through analysis of the stress–strain response under compression loading, measured as the area under the response curve up to the densification strain. Toughness is viewed as a measure of how well the material and cellular design can absorb energy. This property is calculated using MATLAB or Microsoft Excel.
- A, B, m, n – Rusch model parameters. Obtained through fit of experimental data to Rusch model power law functions. This property is calculated using MATLAB or Microsoft Excel.

DOI: 10.1201/9781003299639-3

- C_E, C_y, h, C_D, D – Gibson modified model parameters. Obtained through the fit of experimental data to the piece-wise Gibson modified model equations. This property is calculated using MATLAB or Microsoft Excel.
- A, B, C, m, n – Johnson–Cook (JC) stress parameters. A, B, and n are obtained through analysis of the stress–strain response at low strain rates. C and m are obtained through analysis of the response from the Taylor impact test or Split Hopkinson Pressure Bar (SHPB) test at various strain rates. This property is calculated using MATLAB or Microsoft Excel.
- ν – Poisson's ratio. Obtained through analysis of the stress–strain response under compression loading, the ratio of the transverse strain to the axial strain below the material proportional limit. This property is calculated using digital image correlation.
- D_1, D_2, D_3, D_4, D_5 – JC damage model parameters. D_1, D_2, and D_3 are traditionally obtained through a triaxiality stress test, and D_4 and D_5 are typically obtained from SHPB testing. Another method that has proven successful is an inverse Taylor impact test method, where the damage parameters are backed out from a known end state [3,86].
- A, B, C, N – Holmquist–Johnson–Cook (HJC) strength model parameters. These values are taken from known values for concrete [76,79,87].
- D_1, D_2 – HJC damage model parameters. These values are taken from known values for concrete [76,79,87].

3.2   MECHANICAL TESTING

This section outlines the processes used to fabricate all required test specimens and the mechanical tests required to obtain the necessary material and design properties and parameters.

3.2.1   TEST SPECIMEN FABRICATION

Two different test specimen designs were required for mechanical testing: cylindrical specimens and cube specimens. The specimens were fabricated with three-to-one length-to-diameter ratio to comply with American Society for Testing and Materials (ASTM) quasi-static compression testing standards, a one-to-one length-to-diameter ratio for additional quasi-static compression testing and the SHPB testing, and a five-to-one length-to-diameter ratio for impact testing. While the dimensionality changed based on the method of experimentation, the process for creating each of these designs was nearly identical.

The first step of the process entailed creating a rectangular lattice structure of the chosen cellular design. For this research, the MATLAB software package was used to accomplish this step. The cylindrical specimens were trimmed down to the proper shape in a later stage, requiring a different software package. The design parameters were set within the software starting with the desired cell density or the number of cells across the width of the rectangular structure. The cell density was established by the desired number of cells and required specimen width or diameter. Based on the work of Al-Ketan et al., a minimum cell density of six was chosen [20] for the initial testing. Next, the width of the structure was set and was based on the test specimen requirements. The half-width was also calculated for building the model about the $x - y$ origin. The cell size was set based on the choices made for cell density and specimen width, as it was simply the cell width divided by the cell density. From there, a half-cell size was determined along with the periodicity scaling factor, m, from Equations 1.3–1.6. This scaling factor is a ratio of 2π to the cell size, as shown in Equation 3.1.

$$m = \frac{2\pi}{l} \tag{3.1}$$

The specimen height was determined next. Here was another place that the type of specimen being made impacts the process. For the case of the cylinders, three to five times the width plus a half-cell size was used, which was based on the ASTM standard E9-19 [88], and prior work done with projectile impact testing [89]. For the cubes, the width plus a half-cell size was used. The half-cell size was added to the height to account for the material loss during the removal of the specimen from the build plate in the additive manufacturing (AM) process. Once the rectangle's dimensions were set, a grid was formed using the *meshgrid* function of MATLAB, with the x and y ranges going from negative half-width to positive half-width, and the z range was set from zero to the desired specimen height. A grid was required for the periodic surface equations, as they are functions of x, y, and z. From here, the desired periodic cellular design was chosen from either the triply periodic minimal surface designs, see Equations 1.3–1.6 or variations on those designs, and was coded into MATLAB as a function, where it was evaluated across the three-dimensional grid. The results were then used to create a three-dimensional surface using the *isosurface* function of MATLAB, which outputs a structure file with a list of nodal coordinates and connectivity array for the overlying grid. To be saved properly in stereolithography format referred to as .stl, the isosurface structure file was processed through the *triangulation* function in MATLAB. From here, the file was saved in the .stl format with MATLAB's *stlwrite* function. The .stl file was then posted over to the Rhinoceros 6, referred to as Rhino 6, computer-aided design (CAD) software suite for further manipulation.

Rhino 6 is a non-uniform rational B-splines (NURBS), mathematical representations of three-dimensional geometries, based on a software package that allows for easy manipulation of various CAD and AM file types. In this research, it was used to trim the rectangular structure, if necessary, into a cylindrical form and set the desired surface thickness of the cellular design [90]. Rhino 6 was chosen over similar CAD programs because it can process the intricate nature of the cellular designs on the scale required to develop test specimens. The use of NURBS allowed for smaller file sizes and reduced processing times compared with faceted representation methods. To trim the .stl file, a cylinder of the desired dimensions was created within Rhino 6 as either a mesh or surface. With the imported cellular structure and created cylinder, the *MeshSplit* feature of Rhino6 was used to split the cellular structure along the intersections with the cylinder. The section that was outside of the cylinder was discarded, and a cylinder of the desired dimensions incorporating the cellular design remained. The final action required in the specimen design step was to set the desired surface thickness. Here the *OffsetMesh* feature of Rhino6 was used, where the half-thickness value was input as the offset distance with both sides being selected. Now the test specimen was of the proper dimensions with the desired thickness and was imported to the AM software, Materialise Magics.

Materialise Magics, referred to as Magics moving forward, was the AM software used in conjunction with the General Electric (GE) Concept Laser three-dimensional metal printer used by the Air Force Institute of Technology (AFIT). It is capable of reading in .stl files and preparing them for fabrication. First, a build environment was initiated for the machine to be used, and then the .stl file was imported. Once the file was imported into Magics, it was checked for errors, which could consist of inverted face normals, multiple surfaces, etc. If any errors were found, they can be fixed within Magics through an auto-fix feature or user modifications. The final action required before fabrication was to create the common layer interface, cli, file from the part, which outlines the build path strategy for the printer. The build parameters were set between Magics and the Concept Laser M2 Cusing. The settings used during this research were 130 W of laser power, 1,300 mm/s of laser scan speed, 100 μm of laser beam focus, 100 μm of offset spacing, and 40 μm of layer height. The contour scan settings were used for the lattice structures due to the thin surface thicknesses used to obtain the desired relative densities. Once this was complete, the .cli file is transmitted to the printer, and the part was fabricated. Figure 3.1 depicts the GE Concept Laser M2 Cusing machine used at AFIT.

FIGURE 3.1   Air Force Institute of Technology's 3D metal printer.

3.2.2 Compression Test

The majority of required data were obtained through analysis of the stress–strain response curve under compression loading. Compression testing is a relatively well-known technique for determining material properties in the static and quasi-static regimes, with established procedures [88,91]. Generally speaking, the results from compression testing show the best possible mechanical properties that the material can achieve in ideal conditions. The test process consists of subjecting the specimen of interest to an increasing axial compression force, using either displacement or load control, where both force and displacement data are collected for determining the compression mechanical properties of the material or design. Outlined here are the specific compression testing equipment, procedures, and data reduction methods used as part of this research.

3.2.2.1 Test Equipment

All the static and quasi-static compression tests for this research were performed on an MTS Systems Corporation Model 810 Material Test System. The Model 810 is a servo-hydraulic universal testing machine (UTM) that incorporates the Model 318.50 load unit, capable of producing forces up to 500 kN, or 110 kip, in compression. The system in use also includes an MTS 609 Alignment Fixture. The alignment fixture allows for alignment adjustments while the system is under a preload, which reduces the misalignment error due to applying a load to the system. An MTS 661.23C-01 axial force transducer will also be used, which is force rated to 500 kN, 110 kip. The grip assembly used is an MTS 647.50 Vee-Wedge Grip Assemblies. These grip assemblies are side-loaded hydraulic grips that aid in platen alignment through the use of a sting attachment. The sting attachment screws into the bottom of the platens and automatically aligns along the test area center line when clamped into the vee-wedge. The MTS 643.10B Fixed Compression Platens that will be used are made of a case-hardened steel alloy and incorporate etched rings on the test platform to further aid in test specimen alignment. The machine is calibrated annually by MTS service professionals following ASTM E4-16 calibration procedures. The precision of the UTM as configured is 0.001 mm for displacement and 13 N for force. Figure 3.2 depicts the MTS machine used by AFIT.

3.2.2.2 Test Procedures

To achieve consistent and repeatable results, standardized compression test procedures were used. These procedures are presented here.

1. **Specimen measurement**: Before testing, all the specimens were measured using a digital caliper, with an accuracy of 0.01 mm, and a digital scale, accurate to 0.001 g. Measurements included width or diameter, height, cell size, cell wall thickness, and mass.
2. **Equipment setup**: All necessary equipment was checked and put in place before test initiation. A high-definition digital camera was used to document testing and for digital image correlation of data. A light panel was used to ensure adequate lighting of the specimen during testing. The hydraulic

FIGURE 3.2   Air Force Institute of Technology's MTS universal testing machine.

    pump for the UTM was turned on, and the computer that directs the test program and records data was turned on.

3. **Test program setup**: The software used to run the UTM and record the force and displacement data was the MTS 793 software suite. This software allowed the user to program a loading or displacement test sequence, tare the force and displacement outputs for recording, and compile and save the data for further analysis. In this step, a new specimen file was created for each test specimen and named for the cellular design being tested and the cell wall thickness. Testing was accomplished using a displacement control scheme, and the testing speed was set within the program to meet the recommended 0.005 in/in/min static test compression rate from ASTM [88], or as required. For quasi-static testing, this rate was adjusted over a range of displacement rates to account for specimen dimensions and to fill in gaps between the static and dynamic response ranges, but all tests were performed at a constant rate.

4. **Specimen placement**: Once the system was ready to test and record data, the specimen was placed on the lower platen. Extra care was taken to align the specimen in the center of the platen, using the etched rings and additional markings outlining the desired placement location, to ensure that

moments were not generated due to misalignment. The lower platen was raised until the specimen contacts both lower and upper platen with minimal force registration.

5. **Tare data output**: Before initiating the test schedule, the force and displacement indicators were adjusted to zero.

6. **Conduct test**: The digital camera was initiated to take images at a prescribed rate of 5 Hz before commencing the test program, which automatically started recording the data. The test was continued at the set displacement rate until test completion, set as a force limit of 490 kN or maximum displacement reached of 80% compression of the specimen height. The test was also stopped if any abnormal indications were seen in the data or test specimen, such as highly erratic data or buckling behavior.

7. **Save data**: Once the test was completed, the force and displacement data were saved to the computer and verified before testing a new specimen.

8. **Specimen removal**: Following verification, those data were recorded and saved properly, the specimen was removed from the UTM. The platen was slowly lowered to allow for some elastic unloading of the specimen and then dropped to allow adequate space to remove the specimen.

9. **Equipment shutdown**. Following the last test in the session, all the test equipment was shut down, and the test area was cleaned up before departure.

3.2.2.3 Data Reduction

The dimensional and mass measurements were used to determine the density and relative density of the test specimen. The density equation used for the cylindrical specimen is presented in Equation 3.2 and for the cubic specimens in Equation 3.3.

$$\rho^* = \frac{m}{\pi r^2 h} \tag{3.2}$$

$$\rho^* = \frac{m}{w^2 h} \tag{3.3}$$

where m is the specimen's mass, r is the radius of the cylinder, w is the width of the cube, and h is the specimen's height. Once the specimen's density was determined, Equation 1.1 is used to determine the relative density.

Before determining any of the required mechanical properties or model parameters, the force and displacement data were converted into engineering stress and engineering strain. Force was converted to stress by taking the load applied and dividing it by the original cross-sectional area of the test specimen. Equation 3.4 presents the load-stress conversion equation for the cylinder specimens, and Equation 3.5 offers the load-stress conversion equation for the cube specimens. Since this research required the use of equivalent volumetric material properties, the cross-sectional area was the specimen's bounding area.

$$\sigma = \frac{P}{\pi r^2} \tag{3.4}$$

$$\sigma = \frac{d}{w^2} \tag{3.5}$$

The displacement was converted to strain by taking the displacement, which was the change in height, and dividing it by the original height; see Equation 3.6.

$$\varepsilon = \frac{d}{h_0} \tag{3.6}$$

Here d represents the recorded displacement value, and h_0 is the original specimen height. Once the stress and strain values were acquired, they were plotted as the stress–strain response of the specimen, with stress on the ordinate axis and strain on the abscissa. From this response curve, the following material properties were obtained: elastic modulus, 0.2% offset yield strength, densification strain, and toughness. Through digital image correlation, the Poisson's ratio for the specimens was determined. The Rusch model and Gibson modified model parameters were obtained through the methods described in Section 2.3. Three of the JC flow stress parameters could also be obtained from these data with some additional manipulation.

3.2.2.3.1 Elastic Modulus
As mentioned, the elastic modulus is the slope of the linear elastic region of the stress–strain response. To determine its value, a lower and upper point of the region were determined through analysis of the response. The equation for determining the elastic modulus is presented as Equation 3.7.

$$E = \frac{\sigma_u - \sigma_l}{\varepsilon_u - \varepsilon_l} \tag{3.7}$$

Here the subscript u represents the upper point, and l represents the lower point of the linear elastic region. See Figure 3.3.

3.2.2.3.2 0.2% Offset Yield Strength
The modulus of elasticity must be determined before calculating the 0.2% offset yield strength. On the stress–strain response diagram, a line parallel to the linear elastic region, slope equal to E, was plotted where it crosses the abscissa at 0.2%. The point where this line intersects the stress–strain response was the 0.2% offset yield strength; see Figure 3.4.

3.2.2.3.3 Densification Strain
The densification strain was found in a similar matter to the yield strength. On the stress–strain response diagram, a line parallel to the densification region was plotted, and where it crosses the abscissa was the densification strain; see Figure 3.5.

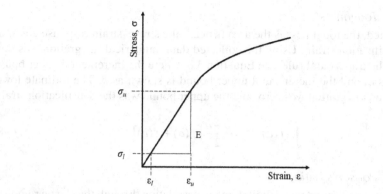

FIGURE 3.3 Modulus of elasticity.

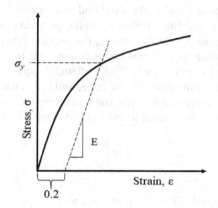

FIGURE 3.4 0.2% offset yield strength.

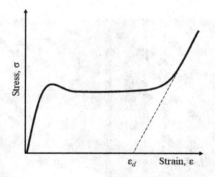

FIGURE 3.5 Densification strain.

3.2.2.3.4 Toughness

As mentioned, the toughness is the area beneath the stress–strain response curve up to the densification strain. Using the collected data, numerical integration was possible with the trapezoidal rule, see Equation 3.8, where the incremental lower bound is shown as *a*, and the incremental upper bound is shown as *b*. The ultimate lower bound for the integration was zero, and the upper bound was the densification strain.

$$\int_a^b f(\varepsilon)d\varepsilon \simeq \frac{b-a}{2}\left[f(a)+f(b)\right] \qquad (3.8)$$

3.2.2.3.5 Poisson's Ratio

Poisson's ratio was found using digital image correlation through the elastic region of the compression response. The initial specimen measurements were correlated to the image at time $t=0$, and the time associated with the end of the linear elastic region was determined during the elastic modulus calculations. Using correlated geometries for the specimen height and width, the axial and transverse strains were determined by comparing the initial specimen geometry to the geometry at the end of the linear elastic regime; see Figure 3.6. In this figure, the image on the left is of a cube test specimen before initiation of uniaxial compression, and the image on the right is at the end of the linear elastic region. The horizontal span in each of the images represents the transverse dimension of the cube, and the vertical spans indicate the axial dimension. From these dimensions, the transverse strain and axial strain were determined. Those values were used to calculate Poisson's ratio; see Equation 3.9.

$$\nu = \frac{\varepsilon_l}{\varepsilon_\alpha} \qquad (3.9)$$

3.2.2.3.6 Johnson–Cook Flow Stress Parameters

Three of the JC flow stress parameters could be determined through axial compression testing, *A, B,* and *n*. These parameters required transforming the engineering strain and stress into true strain and stress; see Equations 3.10 and 3.11. *A* was equal

FIGURE 3.6 Determination of Poisson's ratio through digital image correlation.

to the 0.2% yield stress and was acquired following the same procedures as outlined previously, only in this case using the true stress–strain curve. B and n required isolating the plastic portion of the stress–strain response; see Equations 3.12 and 3.13.

$$\varepsilon_{true} = \ln\left(1 + \varepsilon_{eng}\right) \tag{3.10}$$

$$\sigma_{true} = \sigma_{eng}\left(1 + \varepsilon_{true}\right) \tag{3.11}$$

$$\sigma_{pl} = \sigma_{true} - A \tag{3.12}$$

$$\varepsilon_{pl} = \varepsilon_{true} - \frac{A}{E} \tag{3.13}$$

The plastic stress was plotted versus the plastic strain on a logarithmic scale, where B and n could then be determined through the use of a linear regression fit; see Figure 3.7.

B was calculated as 10 to the power of the intercept of a linear approximation to the curve in Figure 3.7, see Equation 3.14, and the slope of the same linear fit determined n.

$$B = 10^{d} \tag{3.14}$$

3.2.2.3.7 Rusch Model Parameters

The parameters for the Rusch model were determined through regression analysis of the data. The compression response data were split into two sections and then fit with a power law regression to each of the sections. The first section included the linear elastic and plateau regions where A and m were determined. The second section included the origin and densification region where B and n were determined.

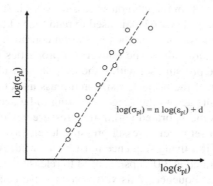

$\log(\sigma_{pl}) = n \log(\varepsilon_{pl}) + d$

FIGURE 3.7 Johnson–Cook flow stress static parameter determination.

The Rusch model was initially taken as the sum of the two power laws, but the coefficients were adjusted to account for the summation of the two terms through an absolute difference fit. The absolute difference technique was chosen against the least-squares fit approach so that a single outlier would not drastically influence the fit of the model.

3.2.2.3.8 Gibson Modified Model Parameters

Similar to the technique used for the Rusch model, the Gibson modified model parameters were determined through regression analysis. In this case, the data were split into three regions with the appropriate regression fit to each section. The first section was the linear elastic region, where E was determined through a linear regression, and C_E was determined by Equation 2.67. The second section, which was the plateau stress region starting at the yield strength, was also fitted with a linear regression. h was determined from the slope of the linear fit curve, and C_y was determined using the 0.2% offset yield strength with Equation 2.70. The final section represented the densification region, which was fitted with a power law regression. From the regression coefficients, the values for D and m were determined through Equation 2.68.

3.2.3 TAYLOR IMPACT TEST

The dynamic material properties and parameters can be obtained using either the SHPB test or Taylor impact test. Due to the facilities available at Wright-Patterson Air Force Base, the Taylor impact test was chosen as the primary means for determining dynamic material parameters.

3.2.3.1 Test Equipment

The ballistic range used by AFIT consists of a pneumatic cannon, which incorporates a pressure vessel, pressure release valve, and barrel. The cannon is mounted to a polycarbonate and aluminum frame that also supports a target mounting platform, projectile backstop, and fragment shielding. The pressure vessel is filled from a standard industrial nitrogen tank, which holds 330 ft³ of nitrogen at 2,500 psi. A Marotta MV74 solenoid valve, rated to 1,800 psi, is used to fill the cannon's pressure vessel. A Sensotec PPG/E981-05-01 pressure transducer, rated to 2,500 psig, provides a readout of the pressure within the cannon's pressure vessel. A National Instruments data acquisition system, NI USB-6210, is used to read the voltages from the pressure transducer, and LabView software on the test station computer converts the voltages to the appropriate pressure value. The conversion properties can be adjusted for a change in gas type. The pressure set within the cannon pressure vessel is correlated to the impact velocities of the projectile size and mass used during this research to achieve repeatable results across the range of testing to be accomplished. A second Marotta MV74 valve is incorporated within the pressure vessel as a pressure release value, which is used to set precise vessel pressure levels or release all the nitrogen from within the vessel if a firing sequence must be aborted. A Circle Seal Controls SV430 solenoid valve, rated at 3,000 psi, is used to release the gas from the pressure vessel during the firing sequence. This valve has a valve coefficient of 0.80 and a stroke length of 0.032 in. The cannon's barrel is made from a smooth-bore steel tube with a half-inch inner diameter. All the valves within the cannon are wired to a 24V power supply with an incorporated electronic safety. An oiled cotton shooting patch will seal the projectile in the barrel not letting the gas leak out around the projectile during the firing sequence. This sealing process is done to ensure consistent and repeatable test conditions.

Two methods were used to determine the impact velocity, ballistic chronographs, and digital image correlation. The current setup included two Caldwell Ballistic Chronographs, calibrated to read within 0.25%. This means that for a 300 m/s shot, there would be no greater than 0.75 m/s of error in the reported velocity. The chronographs incorporate an infrared light attachment, which is used to prevent errors due to the range lighting conditions. The range also incorporated a Phantom v12.1 high-speed camera for use in velocity digital image correlation and recording the impact characteristics of both the projectile and target. Under normal operations, the camera was set with a frame rate of 28,000 frames per second, with a pixel window to frame the dynamic impact zone. For digital image correlation, the camera's pixels were scaled to a measuring device or system within proximity to the projectile's flight path. From this, a correlation of distance per pixel was determined, and the velocity of the projectile was then calculated by dividing the distance traveled by the time interval of the frame rate.

3.2.3.2 Test Procedures

As with the compression test, standardized impact testing procedures were used to achieve consistent and repeatable test results. These procedures are outlined here.

1. **Specimen measurement**: Before testing, the specimens were measured using a digital caliper, with an accuracy of 0.01 mm, and a digital scale, accurate to 0.01 g. Measurements included projectile diameter, length, and mass.
2. **Equipment setup**: All necessary equipment was checked and placed before test initiation. The nitrogen tank pressure was checked to ensure that there is adequate gas for the test phase. The camera was aligned with the impact zone and correlated with a known measurement system before testing. Chronographs were turned on, and all safety equipment was placed in its appropriate location.
   The lighting was adjusted to ensure uniform lighting in the test range, and the computer used to set the cannon's pressure level was turned on.
3. **Test program setup**: As mentioned, LabView was used to convert the pressure transducer voltages into pressures. This program was started, and the initial readout from the system confirmed at 0 psi after the pressure vessel was purged using the pressure release valve.
4. **Projectile placement**: Once the system was ready to test and record data, the range forward of the cannon was cleared before placing the projectile. Next, the projectile was checked for fit confirmation within the barrel and then wrapped in an oiled shooting patch before being placed in the barrel. Care was taken when sliding the projectile down the barrel to ensure that no undue friction or scrapping was observed.

5. **System initialization**: After the projectile was placed within the cannon barrel, the system was prepared for testing. The range was again cleared forward of the barrel, and all valves were checked closed. The pressure vessel was charged to the desired pressure level, which was determined, in part, through an initial correlation of pressure to velocity for the given size of the projectiles used and previous testing.
6. **Conduct test**: The pressure vessel reading was recorded, and the data acquisition system confirmed operational. Protective shielding was confirmed in place, and the range was cleared before initialing the firing sequence. Once the area was confirmed clear, the firing solenoid was activated, and the cannon fired.
7. **Clear range**: Once the firing sequence was completed, the vessel pressure was confirmed at 0 psi, and the vessel pressure release valve was set to open. The protective shielding was removed from the impact area, and the projectile was removed. Once the impact area was cleared of any additional debris, the protective shield was returned to its position.
8. **Save data**: After the range was cleared and safe, the data from the chronographs were recorded, and the data from the high-speed camera were saved before the next test run.
9. **Equipment shutdown**: After the last test in the session, all the test equipment was shut down, and the test area was cleaned up before departure.

3.2.3.3 Data Reduction

As was done for compression testing, the physical measurements of the projectiles were used to determine the density and relative density of each specimen. This allowed for the determination of the relationship between the relative density and dynamic material properties and comparison across the different tested designs.

3.2.3.3.1 Pressure–Velocity Correlation

The first data that needed to be acquired and reduced were the correlation between pneumatic cannon pressure level and impact velocity. Pressure levels were set at 600, 800, 1,000, and 1,200 psi and the corresponding impact velocity recorded. Four projectiles were fired at each pressure level, corresponding to a solid Inconel 718 (IN718) projectile, primitive design, diamond design, and I-WP design. The projectiles had surface thicknesses adjusted to set a common projectile mass. The impact velocity was plotted against the pressure vessel level and a regression fitted to the data to determine the relationship between the two variables. Having a defined pressure–velocity relationship allowed for specific impact velocities and strain rates to be hit during subsequent testing.

3.2.3.3.2 Johnson–Cook Flow Stress Parameters

Under dynamic impact, a projectile tends to maintain its volume; therefore, length and diameter were considered independent measurements, whereas the bulge was regarded as the dimension that conserves the projectile's volume [3]. This left two-cylinder dimensions to be used to solve for the two dynamic JC flow stress parameters.

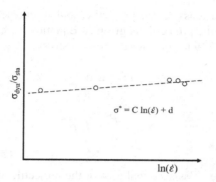

FIGURE 3.8 Johnson–Cook flow stress–strain rate parameter determination.

Of the two remaining JC flow stress parameters, C and m, the value for the strain rate sensitivity parameter, C, was found next. For this, a plot was developed relating the dynamic to static stress ratio and strain rate. Some of these data came from static and quasi-static compression testing following the procedures outlined in the previous section over various strain rates. The high strain rate data came from the impact testing. Using the low-end strain rates acquired from the laboratory compression testing, an initial regression was accomplished. A starting value for C was determined based on the slope coefficient of the linear regression fit; see Figure 3.8. In this case, the JC flow stress parameter C would be equal to the regression coefficient a.

The impact strain rates were determined through the use of the high-speed camera and measuring the projectiles before and after impact, thereby noting the initial length and change in length along with the impact loading time. The compressive stress of impact was approximated using a momentum balance. First, from the analysis of dynamics, the longitudinal speed of sound within a material is shown as Equation 3.15 [11].

$$c = \sqrt{\frac{E}{\rho}}$$
(3.15)

where c is the material speed of sound, E is the material modulus of elasticity, and ρ is the material density. If L gives the initial length of the cylinder, then the contact time of the impact is given as Equation 3.16.

$$t_c = \frac{2L}{\sqrt{\dfrac{E}{\rho}}}$$
(3.16)

Noting that the impulse of the impact force is equal to the projectile's change in momentum, an expression can be developed to balance momentum during impact, see Equation 3.17.

$$F\frac{2L}{\sqrt{\dfrac{E}{\rho}}} = 2mv$$
(3.17)

Here m represents the mass of the projectile, and v represents the impact velocity. Since the mass of the projectile is given as Equation 3.18, Equation 3.17 can be rewritten as Equation 3.19 to solve for the impact stress.

$$m = AL\rho \tag{3.18}$$

$$\sigma_{imp} = \frac{F}{A} = \sqrt{E\rho v} \tag{3.19}$$

where A represents the cross-sectional area of the projectile. Next, an iterative optimization loop was performed to adjust the initial approximation of C to match the impact testing stress results. While there was a range of temperatures that the material was subjected to in the projectile under various impact loads, it was not feasible to determine the exact temperature during impact. Therefore, a finite element analysis (FEA) optimization method was used to determine the value for m. The initial value for m was set to 1.0, which assumed a linear thermal softening. Then through iterations of impact simulations, the value of m was adjusted to reduce the error between computational and experimental results regarding the projectile deformation. A technique by Johnson and Holmquist [3] was used as the minimizing function; see Equation 3.20.

$$\overline{\Delta} = \frac{1}{3}\left[\frac{|\Delta L|}{L_{test}} + \frac{|\Delta D|}{D_{test}} + \frac{|\Delta W|}{W_{test}}\right] \tag{3.20}$$

Here L, D, and W represent measurements of the projectile post-impact. L_{test} is the projectile length after impact, D_{test} is the post-impact diameter of the projectile, and W_{test} is the cross-sectional diameter of a reference location 20% of the projectile's original length back from the impact face, also called the bulge. The Δ terms are the difference between the computational results and the experimental measurements. These dimensions and locations are shown in Figure 3.9.

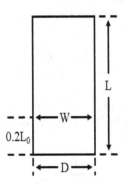

FIGURE 3.9 Impact test damaged projectile measurements.

3.2.4   SPLIT HOPKINSON PRESSURE BAR TEST RESULTS

As a supplement to the Taylor impact data, SHPB testing was accomplished through the University of Alabama.

3.2.4.1   Test Equipment

SHPH components are a popular method for experimentation at high strain rates, from 200 sec^{-1}up to 10^4 sec^{-1}, The components used for these experiments consisted of incident and transmission bars with lengths of 182.25 cm, multiple striker bars with lengths of 29.21 and 44.5 cm depending on the desired strain rate, a shock absorber offset from the end of the transmission bar, and a gas gun to launch the striker bar. All bars had a diameter of 38 mm and were made from 350 C maraging tool steel. Linear strain gages are attached to the incident and transmission bars of 90.17 cm sample–bar interface. A depiction of the test setup is shown in Figure 3.10.

3.2.4.2   Test Procedures

During an experiment, the striker bar is fired from the gas gun, at a velocity between 5 and 35 m/s, toward the open end of the incident bar. This impact causes a stress wave to propagate down the incident bar, passing over the incident bar strain gage, until it reaches the incident bar-sample interface. This wave is referred to as the incident wave. At this point, the incident wave splits, with part being transmitted into the sample, and the rest is reflected. The two stress waves will now be referred to as the transmission wave and reflected wave. The reflected wave will travel back along the incident bar and pass over the incident bar strain gage. Thus, we have two pulses that are recorded by the incident bar strain gage, an incident pulse and a reflected pulse. The transmission wave will travel through the sample and then onto

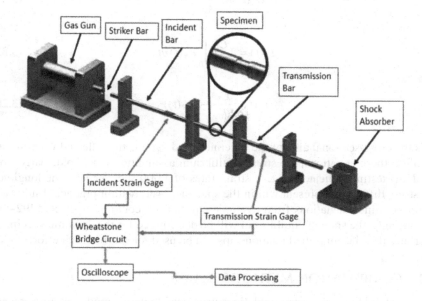

FIGURE 3.10   Split Hopkinson Pressure Bar compression system.

the transmission bar and pass over the transmission bar strain gage. The voltages recorded by the strain gages are converted into the incident strain, ε_l, the reflected strain, ε_R, and the transmitted strain, ε_T. Based on these strain measurements, the geometry and material properties of the bars, and assuming the sample is deformed uniformly, the stress, strain, and strain rate of the sample can be determined. The following equations are used to evaluate the stress, strain, and strain rate, respectively [92]:

$$\sigma = \frac{F_1 + F_2}{2A_b} = \frac{1}{2}E_b\left(\varepsilon_l + \varepsilon_R + \varepsilon_T\right) \tag{3.21}$$

$$\varepsilon = \frac{u_1 - u_2}{l_0} = \frac{C_b}{l_0}\int_0^t \left(\varepsilon_i - \varepsilon_R - \varepsilon_T\right)dt \tag{3.22}$$

$$\dot{\varepsilon} = \frac{v_1 - v_2}{l_0} = \frac{C_b}{l_0}\left(\varepsilon_i - \varepsilon_R - \varepsilon_T\right) \tag{3.23}$$

Here, F_1 and F_2 represent the force acting at the front face and back face of the specimen, respectively. A_b, E_b, and C_b are the cross-sectional area, modulus of elasticity, and wave velocity of the incident and transmission bars, respectively, and l_0 is the initial length of the specimen. u is the displacement, and v is the velocity, where $v = du/dt = C_b\varepsilon$. Furthermore, if the specimen maintains the condition of dynamic equilibrium, where $F_1 = F_2$ or $\varepsilon_l(t) + \varepsilon_R(t) = \varepsilon_T(t)$, these equations can be simplified to

$$\sigma_s(t) = \frac{E_b A_b}{A_s}\varepsilon_T(t) \tag{3.24}$$

$$\varepsilon_s(t) = 2\frac{C_b}{l_0}\int_0^t \varepsilon_R(t)\partial t \tag{3.25}$$

$$\dot{\varepsilon}_s(t) = \frac{C_b}{l_0}\left(-2\varepsilon_R(t)\right) \tag{3.26}$$

A_s is the cross-sectional area of the specimen, and $\varepsilon_R(t)$ is the reflected wave at time t. To use these equations, the stress equilibrium assumption must hold. Early work in SHPB testing indicated that, at strain rates of 1,000 sec^{-1} or less, the longitudinal stress differences present within the specimen were within practical limits, and therefore minor deviations from dynamic equilibrium could be neglected [92–94]. Accordingly, the state of equilibrium will be determined for the dynamic specimens to ensure that the simplified relationships can be used for this research effort.

3.3 COMPUTATION METHODS

Computational methods and simulations were used to derive model parameters, and A_s form a basis of evaluation for the different periodic cellular designs under impact loading. First, the remaining model parameters will be discussed, followed by a description of the computational impact model, and analysis methods.

3.3.1 Johnson–Cook Damage Model Parameters

The JC damage model parameters were used within the impact simulations as part of the viscoplastic effects modeling and element or particle failure assessments. These parameters are typically found through a combination of triaxiality and SHPB testing. However, in this research, an indirect, inverse method was used to determine the parameters following the work accomplished by Banerjee [86]. In this method, the Abaqus FEA program performed numerical simulations of the compression and impact tests using the elastoplastic material model. First, the material model parameters were adjusted to match and validated against the experimental uniaxial compression test data. Once that was complete, the simulation was used to determine the effect of triaxiality ratio on the strain response with constant strain rate and temperature, which provided the data necessary to establish the first three JC damage model parameters, D_1–D_3. The term *normalized* failure *strain* is used when discussing the plots required to determine the damage parameters. Normalized failure strain is simply the resultant strain from only varying one of the three terms within the overall JC fracture strain, Equation 2.57. The normalized failure strain was plotted against the triaxiality ratio, and a power law regression was performed to determine D_1, D_2, and D_3 from the first term in Equation 2.57. See Figure 3.11.

The next parameter, D_4, was determined through analysis and simulation of impact data. Matching the responses obtained through experimentation and introducing an additional variation of impact strain rate through simulation, all, while maintaining a constant temperature within the model, provided the necessary data to establish a strain rate to failure relationship that forms the basis for D_4. In this case, the normalized failure strain was plotted against the natural log of the strain rate and fit with a linear regression. The slope of the regression line provides D_4; see Figure 3.12.

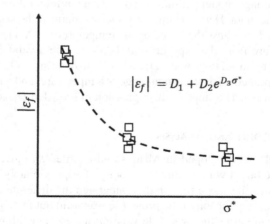

$$|\varepsilon_f| = D_1 + D_2 e^{D_3 \sigma^*}$$

FIGURE 3.11   Johnson–Cook damage model triaxiality parameters.

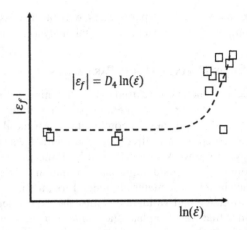

FIGURE 3.12   Johnson–Cook damage model strain rate parameter.

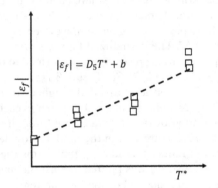

FIGURE 3.13   Johnson–Cook damage model temperature sensitivity parameter.

The final JC damage model parameter, D_5, was again found through analysis and simulation of impact data. Here, the impact velocity remained the same, and the temperature was varied to adjust the homologous temperature, T^*. The first evaluation used the temperature from the experimental data to create a validation point. Then the temperature was varied to provide a range of T^* to determine D_5. The normalized failure strain was plotted against the homologous temperature, and a linear regression was used to fit the data. The slope of the regression fit provided D_5; see Figure 3.13.

3.3.2   Impact Model and Analysis

The impact model was developed in Abaqus and entailed an ogive nose projectile striking a concrete target with an impact velocity of approximately 250 m/s. Within the nose of the projectile was a section that integrated the different periodic cellular designs. The remaining portion of the projectile was solid but included a channel for the powder to escape after the projectile was fabricated. This allowed for a direct comparison of how the different designs performed under the same conditions while reducing the number of free variables to isolate an optimal design for this condition. Each of the model parts, as well as the overall assembly environment, will be discussed in more detail.

3.3.2.1 Projectile

As a means for comparison between the different cellular designs under consideration as part of this research, a common projectile was used to evaluate the performance of the designs under a high strain rate impact with a concrete target. The projectile was 12.7 mm in diameter and 63.5 mm in length, with an internal section that had a maximum diameter of 11.7 mm and a length of 25.4 mm that contained the lattice structure, all made of IN718. Figure 3.14 depicts the projectile with the internal cellular lattice core represented by the primitive design.

To carry out the evaluation, three performance measures were used, each of them further outlined here.

3.3.2.1.1 Measures of Performance

- **Penetration depth**: The distance that the nose of the projectile penetrated within the concrete target was the first measure of performance. This is a commonly used measurement of impact performance, as it is one of the primary considerations of impact modeling and survivability. The measurement was taken from the center of contact on the impact face of the target and measured as a straight line from that face to the point where the projectile tip ceased forward movement.
- **Projectile deformation**: The second measure of performance was the amount of deformation sustained by the projectile under the impact conditions. This was primarily a measurement of the utility of cellular inclusion within a projectile. If the projectile maintained its designed shape further into the target, it would impart the intended damage to the target. This was also an indirect measurement of how well the cellular design inclusion absorbed and controlled energy flow, through the form of a stress wave, during an impact event. The measurements used here were the overall change in projectile length, change in the diameter at the mid-point of the projectile, and change in the bulge diameter.

FIGURE 3.14 Projectile for modeling and simulation.

- **Energy absorption**: The final measurement of performance was the ability of the inclusion region to absorb the energy imparted within the projectile during impact. This entailed tracing the stress wave as it propagated through the projectile and evaluating the stress intensity change and wave velocity change across the inclusion region. For this measurement, common analysis paths were created along the center line and at the edges of the projectile, where the stress levels and wave velocity were traced.

The projectile was modeled using traditional Lagrangian FEA implementation techniques. The material properties for the projectile and inclusion were based on the testing done as part of this research.

3.3.2.2 Target
A common concrete target was used for the evaluation of each of the projectiles. It was a rectangle with a 101.6 mm by 203.2 mm of cross-section face and with a length of 304.8 mm. The material properties and HJC model parameters were taken from previous literature [87], as shown in Table 3.1.

The concrete target was modeled using smoothed particle hydrodynamics (SPH) FEA implementation and techniques. This was done due to the highly dynamic nature of brittle failure at high-impact velocities. The cubic spline kernel function was used with three-dimensional stress particles, and ghost particles were incorporated to control the SPH–Lagrangian boundary conditions.

3.3.2.3 Model Assembly
Within   starting at the initial contact between the projectile and target. The projectile was placed with its center line perpendicular to the target face, zero incidence

TABLE 3.1
HJC Model's Concrete Material Properties

Material Property	Units	Value
Density, ρ	$\dfrac{\text{kg}}{\text{mm}^3}$	2.24×10^{-6}
Shear modulus, G	GPa	13.567
HEL	GPa	2.79
PHEL	GPa	1.46
A	-	0.75
B	-	1.65
C	-	0.007
N	-	0.76
$\dot{\varepsilon}_0$	$\dfrac{1}{\text{ms}}$	0.001
$\sigma_{f_{max}}^{*}$	-	0.048
T	GPa	0.004
D_1	-	0.03
D_2	-	1

angle, and the velocity was set to 250 m/s for impact. The two sections within the projectile were merged to create one monolithic projectile with different material properties. Cohesive elements were not used as the intention was that the projectile would be manufactured additively, and therefore no bonding would be required. The target had boundary conditions applied along its edges, not allowing movement or rotation as if it were fixed to the test stand. The model was initially set to run for 2.0 ms and then adjusted to account for any projectile residual velocity to ensure that the projectile had reached its maximum penetration depth and the simulation was complete. Simulations were run with the full model to collect data on the total system under investigation. In addition, reduced-scale model simulations were developed based on the conditions of symmetry within the model. These models split the full-scale model in half along the vertical center line, which allowed for reduced computation time and improved visualization within the lattice inclusion.

3.4 TOPOLOGY OPTIMIZATION

The following section will introduce the mathematical details of the topology optimization problem and then explore the relationship between the micro and macro scales, discussing potential methods for bridging the gap between them. Two widely used optimization methodologies will be presented in greater detail: bi-directional evolutionary structural optimization (BESO) and solid isotropic material with penalization (SIMP).

3.4.1 OPTIMIZATION OVERVIEW

Structural optimization is a mathematical approach of fundamental interest to the engineering field due to its potential for future growth and development. At its core, optimization is concerned with maximizing, or minimizing, certain properties under specific loading conditions and constraints on the structure. The subject of structural optimization is broken into three separate but related sub-domains: size optimization, shape optimization, and topology optimization. In engineering applications, these three sub-domains are often treated separately, as the size and shape of a structure are typically determined by utility and requirements, having a minimal design space available to consider. Thus, topology optimization is of utmost interest when attempting to construct the most effective design of structures.

While the shape and size of a given structural design are well understood, the consideration of topology in the design process is much more esoteric. Geometrically, topology is concerned with the arrangement of the material itself within the given structural shape and size. As the goal is generally to maximize a specific structural or mechanical quality while minimizing the amount of material required, a firm understanding of the applicable principles of topology optimization is necessary.

The first step in preparing a structure for a topology optimization analysis is to understand the design space. That is, the nature of design decisions made throughout the topology optimization must be formalized. This aim is achieved by performing an FEA on the structure. Within the prescribed shape and size, the continuum design space is partitioned into a mesh of discrete elements of finite size. Then a structural analysis is carried out on the elements. This type of analysis is widely used within the field of mechanics, and thus a variety of tools and software packages exist for FEA, such as Abaqus [95] and Fusion 360 [96]. This formulation naturally allows for design decisions based on the property to be optimized and the constraints placed on the structure. That is, for each finite element within the mesh, the determination is made on whether the material should exist as part of the solution or if the mesh volume should be left void.

The optimization problem can be formulated with the finite element method because it allows incorporation of design decisions for a structure's topology. Canonically, any optimization problem can be expressed through well-defined design variables, an objective function, and constraints on the size and shape of the structure. The generalized optimization problem is expressed below, where $f(\mathbf{x})$ is the objective function to be minimized, $\mathbf{x}$ is the vector of design variables, and $\mathbf{C}_j$ is the jth constraint to be satisfied by the design variables.

$$\text{Minimize } f(\mathbf{z}) \text{ subject to :}$$

$$\mathbf{C}_j$$

$$\mathbf{z}_j = 0 \text{ or } 1$$

For a structure subject to extreme loading or extreme boundary conditions, it is of particular interest to explore the microscale effects of the material by which the structure is composed. Historically, topology optimization has been performed under the assumption of linear elastic structures with negligible microscale effects, and for larger homogeneous structures undergoing standard loading, it was an effective assumption that permitted more easily calculable solutions. However, with more strenuous engineering requirements, such as a complex starting design like an open cell lattice, it is necessary to consider the microstructure of the material composing the structure. At a fundamental, physical level, every material can be characterized by the structural formation of its atoms. Metals are characterized by the formation of periodic crystal structures composed of their respective atoms. These structures then form the grain microstructure of the metal. The governing behavior of this phenomenon is well understood through materials modeling in physics. As the size of the elastic structure approaches the scale of its constitutive microstructure, the behavior of the microstructure grows in significance.

3.4.2 MULTISCALE MODEL

To formulate the topology optimization problem while accounting for both the macroscale and microscale effects within the structure, the first-order homogenization method, FE$_2$, will be considered [97]. This method assumes a distinct separation of the treatment of the two spatial scales, including their periodicity assumptions. The procedure seeks to attain macroscale equilibrium within the structure given an initial loading condition. At the macroscale, the structure is assumed to be homogeneous, wherein each discrete finite element maintains unknown structural characteristics to be determined by the constitutive microstructure. Figure 3.15 provides a visual

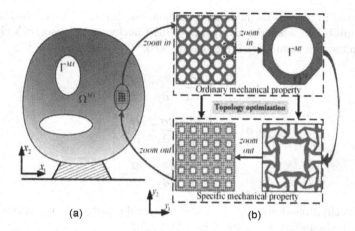

FIGURE 3.15 Relationship between macrostructure and microscale material properties
[98]: (a) micro-structured material and (b) material cell.

representation for the characterization of the structural macroscale and material
microscale.

For a given material, the potential relation as a function of microstructure strain is
assumed to be known, where the microstructure stress is the partial derivative of the
potential function with respect to strain:

$$\sigma(\varepsilon) = \frac{\partial \omega(\varepsilon)}{\partial \varepsilon} \qquad (3.27)$$

Next, periodic boundary conditions for each structural element within the finite element framework are established. Iteratively, given the initial loading conditions, the
microstructure strain at each discrete material point within the discrete structural
point is set equal to the structural average strain value, or mean strain, represented
by $\bar{\varepsilon}$ [99]. Letting $\mathbf{X}$ denote the spatial position of the macroscale element, and letting
x denotes the position of the microscale element within the structural element, the
following relation is established in Equation 3.28:

$$\varepsilon(\mathbf{X}, x) = \bar{\varepsilon}(\mathbf{X}) \qquad (3.28)$$

Using this relation, the microstructure stress, $(\mathbf{X}; x)$, is then calculated using Equation
3.29 [99]. The macroscale stress at each structural element is then found by taking
the mean stress, across the material elements through means of volume averaging
across the interfacing elements:

$$\sigma(\mathbf{X}) = \bar{\sigma}(\mathbf{X}, x) \qquad (3.29)$$

This method assumes small displacements within the macrostructure, and proceeds
to calculate the tangent stiffness tensor and subsequently update the macroscale
displacement using a numerical technique, such as the Newton–Raphson Method,

which is used in the FE$_2$ methodology. The macrostructure is determined to be in a state of equilibrium when the divergence of the Cauchy stress tensor, $\sigma(\mathbf{X})$ is zero at every structural element shown in Equation 3.30,

$$\nabla \cdot \sigma(\mathbf{X}) = \begin{bmatrix} \dfrac{\partial \sigma_{xx}}{\partial x} & \dfrac{\partial \sigma_{yx}}{\partial y} & \dfrac{\partial \sigma_{zx}}{\partial z} \\[2ex] \dfrac{\partial \sigma_{xy}}{\partial x} & \dfrac{\partial \sigma_{yy}}{\partial y} & \dfrac{\partial \sigma_{zy}}{\partial z} \\[2ex] \dfrac{\partial \sigma_{xz}}{\partial x} & \dfrac{\partial \sigma_{yz}}{\partial y} & \dfrac{\partial \sigma_{zz}}{\partial z} \end{bmatrix} = 0 \qquad (3.30)$$

and the Cauchy stress tensor is equal to the external Cauchy traction vector, **t**, along the structural boundary as shown in Equation 3.31:

$$\mathbf{t} = \sigma \cdot n \qquad (3.31)$$

In essence, this method establishes an equilibrium between the structural macroscale, characterized by the potential energy of the external forcing, and material microscale effects, described by the potential energy of the internal strain, through the implementation of periodic boundary conditions between the two scales, and iteratively updating the macroscale displacements through non-linear elastic analysis while checking that necessary equilibrium conditions are met. Figure 3.16 provides a visual depiction of this iteration process.

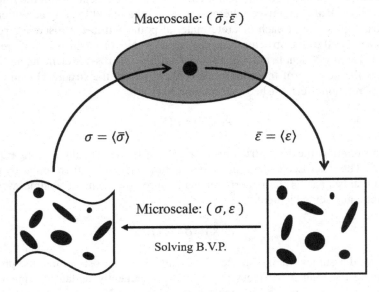

Macroscale: $(\bar{\sigma}, \bar{\varepsilon})$

$\sigma = \langle \bar{\sigma} \rangle$ $\bar{\varepsilon} = \langle \varepsilon \rangle$

Microscale: (σ, ε)

Solving B.V.P.

FIGURE 3.16   Iterative solution procedure of the FE$_2$ methodology.

As might be expected, the iterative nature of the FE_2 methodology for resolving macroscale and microscale effects within a structure comes with significant computational costs. The solution requires the storing of microscale data within each macroscale element, which increases computer memory requirements, especially for fine finite element meshes. Furthermore, the computation of the macroscale displacement requires numerical root finding methods, which may require several iterations depending on desired accuracy.

3.4.3 TOPOLOGY OPTIMIZATION METHODS

The conventional optimization problem requires the maximization or minimization of a specific objective function over a set containing design variables subject to a system of constraints that must be met. These constraints are typically set by design limitations or requirements. For linear optimization problems, the solution methods are well established and relatively easy to implement. Such problems are solved using linear programming through the simplex algorithm, which evaluates the objective function while traveling along vertices of the n-dimensional polytope bounded by the permissible region that is established by the constraints on the n decision variables. This procedure assumes a linear relationship between the objective function, system constraints, and design variables. In addition, the procedure requires that the design space be continuous. For the non-linearly elastic topology optimization problem described in this paper, neither of these conditions are met. Therefore, non-linear optimization methods must be used in an attempt to solve the topology optimization problem at hand. By default, discrete topology optimization algorithms within FEA systems have the disadvantage that the product of the optimization is a non-smooth structural geometry. As many engineering applications require smooth geometric shapes, a smoothing procedure has to be performed. Depending on the optimization geometric constraints, the resulting geometry is usually highly organic in shape, requiring a manual process of interpreting and implementing the results into a parameterized model suitable for further computational analysis or production purposes.

In developing the topology optimization model, the first consideration must be determining precisely what is to be optimized within the structure. For many structural applications within the aerospace industry, interest often lies with minimizing structural deformation due to an applied loading profile within the prescribed size and weight constraints. This phenomenon is captured through the stiffness of the structure, which can be thought of as the structure's resistance to deformation at a given applied load. Due to its importance in application, the focus of this chapter will be on the effective maximization of structural stiffness. An equivalent objective to maximizing stiffness is minimizing the mean compliance, or flexibility, of the structure. Mean compliance is defined as (see Equation 3.32):

$$C = \mathbf{f} \cdot \mathbf{u} \tag{3.32}$$

where $\mathbf{f}$ is the applied load vector acting on the structure and $\mathbf{u}$ is the element displacement at the applied load. This objective function is equivalent to minimizing the strain energy of the structure under static loading cases, while neglecting other energy effects such as thermal considerations. As structural elements are added and removed in the pursuit of an optimal topology, the overall system is constrained by the total volume of possible structural elements. That is, the total sum of structural elements must be less than or equal to the total allowable number of structural elements as defined in the finite element discretization of the structure. Additionally, a load equilibrium constraint must be imposed on the structure. The externally applied force must be in equilibrium with the subsequent internal loads experienced within the structure, which can be expressed through the relation of Equation 3.33:

$$f - \sum_{i=1}^{N} x_i \int_{V_i} \mathbf{B}^{\mathrm{T}} \sigma \, d\Omega_i = 0 \qquad (3.33)$$

where x_i is the ith design variable representing the binary existence of an element, $\mathbf{B}$ is the shape function matrix for the element, σ is the Cauchy stress tensor, and Ω_i is the ith structure elemental volume. Combining the objective function with these constraints, the topology optimization problem is obtained:

$$\text{Minimize } C = \mathbf{f} \cdot \mathbf{u}$$

$$\text{Such that :}$$

$$V_{\mathrm{tot}} - \sum_{i=1}^{N} x_i V_i \geq 0$$

$$f - \sum_{i=1}^{N} x_i \int_{V_i} \mathbf{B}^{\mathrm{T}} \sigma \, d\Omega_i = 0$$

$$\text{Where}$$

$$x_i = 0 \text{ or } 1$$

Solving such a non-linear, integer programming problem is highly nontrivial, and must rely on numerical methods to obtain a solution.

3.4.3.1 Bi-directional Evolutionary Structural Optimization

One of the most prominent methods developed to address non-linear, numerical optimization in the past couple of decades is the BESO method. This method is concerned with establishing an algorithm that converges stably toward the optimal solution by simultaneously adding and removing structural elements and evaluating the resulting impact on the objective function. This requires the introduction of an ith elemental sensitivity number, α_i^e which is defined by Equation 3.34:

$$\alpha_i^e = \Delta C_i = \frac{1}{2} u_i^T \mathbf{K}_i u_i \qquad (3.34)$$

where $\mathbf{K}_i$ is the ith elemental stiffness matrix. This number measures the change in the objective function due to a change in displacement from adding or removing a structural element. In the BESO procedure, the sensitivity number is first calculated for each structural element and sorted in size. Additionally, an evolutionary volume ratio (ER) must be predefined, which limits the total possible change in volume between successive steps k in the algorithm (see Equation 3.35):

$$V_{k+1} = V_k(1 \pm ER) \tag{3.35}$$

This parameter is determined through intuition and experience, and must strike a balance between convergence speed and solution stability. It is standard to set $ER = 1\%$ for moderately sized problems. Once each sensitivity parameter is calculated, the algorithm determines which structural elements to add or remove by comparing the previous volume iteration to the current value. Elements with sensitivity numbers greater than the sensitivity value corresponding to the threshold value from the newly calculated volume is added, with the converse being true for low sensitivity numbers. After the addition and removal of elements, the objective function is updated and a relative error is computed to measure the state of convergence:

$$r = \frac{\left| \sum_{i=1}^{M} C_{k-i+1} - \sum_{i=1}^{M} C_{k-M-i+1} \right|}{\left| \sum_{i=1}^{M} C_{k-i+1} \right|} \tag{3.36}$$

where M is usually set to five, to measure the relative error across the last five steps [100]. Once the relative error falls below a chosen threshold, $r \leq \tau$, the algorithm terminates and the optimization problem is solved. It is worth noting that this iterative procedure fails to guarantee solution convergence, and could be susceptible to slow convergence rates and significant fluctuations in objective function evaluations. Furthermore, since each step in the procedure requires the evaluation of the sensitivity number at each structural element, along with sorting the values, this algorithm carries a significant computational burden when tasked with performing topology optimization over large scales and fine element meshes. However, the BESO procedure is a frequently used method of evaluating the topology optimization problem and can easily be implemented in software packages such as MATLAB, Fusion 360, and Abaqus.

3.4.3.2 Solid Isotropic Material and Penalization Method

Another prevalent optimization technique is the SIMP method. This method assumes that each structural element composing the finite element discretization of the structure is made of isotropic materials, where each structural element can have variable density. This formally eliminates the binary design variable restriction that is present in the BESO method, which enables the introduction of solution methods prohibited by the integer requirement. However, the underlying principle behind the design decisions is effectively the same, as the elemental densities are driven by a power law material interpolation scheme to take values close to zero or one. Aside from this fundamental shift, the presentation of the optimization problem is effectively identical:

$$\text{Minimize } C = \mathbf{f} \cdot \mathbf{u}$$

Such that :

$$V_{\text{tot}} - \sum_{i=1}^{N} x_i V_i \geq 0$$

$$f - \sum_{i=1}^{N} x_i \int_{V_i} \mathbf{B}^T \sigma \, d\Omega_i = 0$$

Where

$$0 \leq x_{\min} < x_i \leq 1$$

where $x_{\min}$ is a small number to ensure that no singularities are encountered. Just as in the BESO method, the sensitivity of the objective function to the existence or non-existence of a structural element is computed. Since SIMP assumes continuous design variables, the partial derivative of the objective function can be taken with respect to the design variable at the ith structural element:

$$\frac{\partial c}{\partial x_i} = -p x_i^{p-1} u_i^T K_i^0 u_i \qquad (3.37)$$

where $p \geq 3$ is the penalization factor, which leads to the polarization of the design variables to take on an effective binary value [101]. From this point, the SIMP method diverges notably from the BESO method when updating the continuous design variables. Here, the ith design variable updated at the Kth iteration is found by comparing the $K-1$ iteration's design variable to a numerical parameter found through the application of optimality criteria or conditions [102,103]. In evaluating the performance of the SIMP method, it is useful to note that due to the penalization scheme imposed on the continuous design variables it is possible to arrive at locally optimized solutions while iterating through the solution process. Additionally, there is no guarantee of a convergent solution, much in the same fashion as the BESO method. One distinct advantage the SIMP method maintains over the BESO method is that the convergence of the solution is less dependent on user-selected parameters.

Though not shown here in evaluating the I_WP lattice cell, optimization provided 62.6% reduction in maximum stress and a 30% in volume and mass. The primitive lattice cell showed similar advantage through the optimization.

REFERENCES

3. G. R. Johnson and T. J. Holmquist. Evaluation of cylinder-impact test data for constitutive model constants. *Journal of Applied Physics*, 64(8): 3901–3910, 1988.

11. Michael F. Ashby. *Materials Selection in Mechanical Design*. Elsevier, 3rd edition, 2005.

20. O. Al-Ketan, R. Rowshan, and R. K. Abu Al-Rub. Topology-mechanical property relationship of 3D printed strut, skeletal, and sheet based periodic metallic cellular materials. *Additive Manufacturing*, 19(January): 167–183, 2018.

76. T. J. Holmquist, G. R. Johnson, and W. H. Cook. A computational constitutive model for concrete subjected to large strains, high strain rates, and high pressures. In *Proceedings of the 14th International Symposium on Ballistics*, pages 591–600, Quebec City, 1993.

79. G. M. Ren, H. Wu, Q. Fang, and X. Z. Kong. Parameters of Holmquist–Johnson–Cook model for high-strength concrete-like materials under projectile impact. *International Journal of Protective Structures*, 8(3): 352–367, 2017.

86. A. Banerjee, S. Dhar, S. Acharyya, D. Datta, and N. Nayak. An experimental determination of Johnson Cook material and failure model constants for armor steel. *Applied Mechanics and Materials*, 592–594: 990–995, 2014.

87. W. T. Graves, D. Liu, and A. N. Palazotto. Impact of an additively manufactured projectile. *Journal of Dynamic Behavior of Materials*, 3(3): 362–376, 2017.

88. ASTM International. Standard Test Methods of Compression Testing of Metallic Materials at Room Temperature, 2019.

89. C. Hernandez and A. Maranon. Determination of constitutive parameters from a Taylor test using inverse analysis. *Strain*, 53(6):1–13, 2017.

90. Robert McNeel and Associates. Rhinoceros 6, 2020.

91. C. Lakshmana Rao, V. Narayanamurthy, and K. R. Y. Simha. *Applied Impact Mechanics*. John Wiley & Sons Ltd, Amsterdam, Netherlands, 2016.

92. E. W. Billington and C. Brissenden. Mechanical properties of various polymeric solids tested in compression. *International Journal of Mechanical Sciences*, 13(6): 531–545, 1971.

93. U. S. Lindholm. Some experiments with the split Hopkinson pressure bar. *Journal of the Mechanics and Physics of Solids*, 12(5): 317–335, 1964.

94. U. S. Lindholm and L. M. Yeakley. High strain-rate testing: Tension and compression. *Experimental Mechanics*, 8: 1–9, 1968.

95. Dassault Syst`emes. Abaqus unified finite element analysis software, 2016.

96. Autodesk. Fusion 360 unified computer aided design and computer aided modeling software, 2020.

97. L. Xia. Topology optimization framework for multiscale nonlinear structures. In Liang Xia, editor, *Multiscale Structural Topology Optimization*, Amsterdam, Netherlands, pages 1–19. Elsevier, 2016.

98. J. Gao, H. Li, Z. Luo, L. Gao, and P. Li. Topology optimization of micro-structured materials featured with the specific mechanical properties. *International Journal of Computational Methods*, 17: 1850144, 2018.

99. L. Xia and P. Breitkopf. Multiscale structural topology optimization. *11th World Congress on Structural and Multidisciplinary Optimization*, Sydney, Australia, pages 1–6, 2015.

100. X. Huang, Y. M. Xie, and M. C. Burry. Advantages of bi-directional evolutionary structural optimization (beso) over evolutionary structural optimization (eso). *Advances in Structural Engineering*, 10(6): 727–737, 2007.

101. H. Eschenauer and N. Olhoff. Topology optimization of continuum structures: A review. *Applied Mechanics Reviews*, 54: 331–390, 2001.

102. O. Sigmund and K. Maute. Topology optimization approaches: A comparative review. *Structural and Multidisciplinary Optimization*, 48(6): 1031–1055, 2013.

103. L. Li and K. Khandelwal. Topology optimization of structures with length-scale effects using elasticity with microstructure theory. *Computers & Structures*, 157: 165–177, 2015.

4 Uniaxial Compression of Lattices

Following the path outlined by the research objectives, see Section 1.2, the first step in characterizing the time-dependent behavior of lattices was to determine the quasi-static response of the different lattice cell designs.

4.1 UNIAXIAL COMPRESSION OF CYLINDERS

The first set of uniaxial compression testing was accomplished on the Air Force Institute of Technology (AFIT) Universal Testing Machine (UTM) as described in Section 3.2.2.1 following the process outlined in American Society for Testing and Materials (ASTM) E9-19 and those provided in Section 3.2.2.2. ASTM E9-19 was chosen as it is a standard test method for metallic materials, even though it was not necessarily intended to be used for periodic structures. The medium-length test specimen dimensions and aspects were chosen to determine the general compressive properties [88]. In addition to the information provided by the machine load cell, high-resolution photography with five images taken per minute, and videography were used. The goal of testing was to determine the modulus of elasticity, yield strength, yield point, and plateau strength. Only three of the triply periodic minimal surface (TPMS) designs were fabricated for this stage of testing, the primitive, diamond, and I-WP. The Lidinoid design was not created due to processing errors in the initial specimen development process, which was fixed before the second round of uniaxial testing. Before laboratory testing, the actual relative density was compared with the predicted design relative density. The predicted design relative density was determined using the properties of the specimen as determined by Materialise Magics compared with a solid specimen of the same dimensions. The actual relative density was determined following the procedures presented in Section 3.2.2.3. As shown in Figure 4.1, the actual relative density was higher for each of the designs than predicted. The primitive TPMS structures were 3.1% denser than the computational prediction, whereas the diamond TPMS structures were 2.5% denser, and the I-WP TPMS structure was 5.3% denser than designed.

One of the solid specimens and all the periodic cellular structured cylinders were tested, and all but one of the primitive designs provided usable data. All the specimens exhibited buckling behavior and therefore could not be used in determining the plateau strength, and further analysis was performed to determine if the yield strength and yield point were valid.

Figure 4.2 presents the results for the solid cylinder. From these data, the modulus of elasticity was calculated to be 36.58 GPa, and the 0.2% offset yield was determined to be −598.52 MPa. The jump in the data at −0.054 mm/mm was due to having to

DOI: 10.1201/9781003299639-4

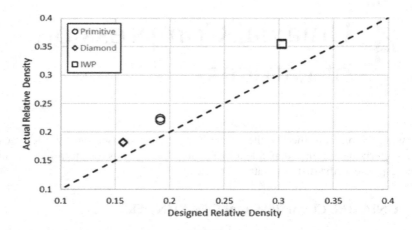

FIGURE 4.1   Comparison of cylinder design relative density.

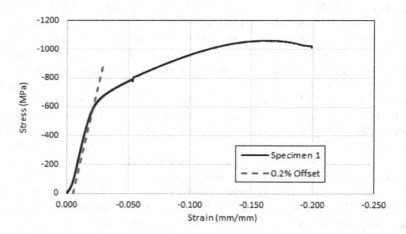

FIGURE 4.2   Uniaxial compression of solid cylinder.

restart the compression program. Overall, the data follow the same curve, so this was not seen as a significant issue as this happened after the cylinder began to buckle.

Figure 4.3 presents the results obtained for the primitive cylinder. From these data, the modulus of elasticity was calculated to be 3.13 GPa, and the 0.2% offset yield was determined to be −38.79 MPa. As mentioned above, only one of the two primitive design cylinders provided usable data. Data from the other cylinder were not able to be retrieved from the computer hooked up to the MTS Systems Corporation machine.

Figure 4.4 presents the results obtained for the diamond cylinder. In general, there was good agreement between the two specimens, especially in the elastic region. From these data, the modulus of elasticity was calculated to be 3.07 GPa, and the 0.2% offset yield was determined to be −36.08 MPa.

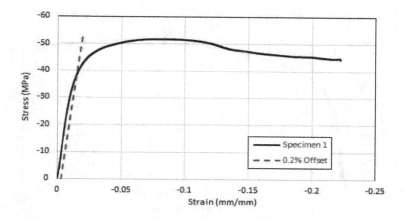

FIGURE 4.3 Uniaxial compression of primitive cell cylinders.

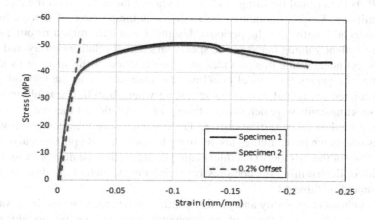

FIGURE 4.4 Uniaxial compression of diamond cell cylinders.

Figure 4.5 presents the results obtained for the I-WP cylinder. Again, the two test specimens showed good agreement throughout the range of data collected and were nearly identical in the elastic region. From these data, the modulus of elasticity was calculated to be 5.64 GPa, and the 0.2% offset yield was determined to be −75.54 MPa.

For each of the specimen designs, a buckling stress was determined using the Euler equation; see Equation 4.1 [105].

$$P_{cr} = \frac{\pi^2 EI}{L^2} \tag{4.1}$$

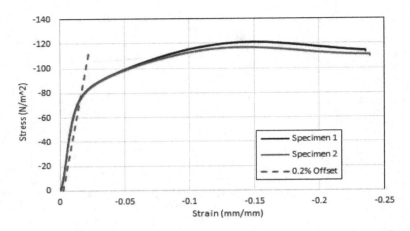

FIGURE 4.5   Uniaxial compression of I-WP cell cylinder.

where P_{cr} is the critical buckling load, n is a scaling factor based on the end boundary conditions, E is the modulus of elasticity, I is the moment of inertia, and L is
the cylindrical length. For the periodic designs, l was determined through the use
of an equivalent volume between the specimen's actual relative density and that of
a hollow cylinder. While this provides an approximation of the specimen's moment
of inertia, the values for critical buckling load obtained using the method slightly
underpredicted the actual load point in testing where buckling was observed. The
actual buckling loads were determined through a correlation of the load data and the
photographic data taken during testing. These values were then compared with the
yield results shown in the figures above. In each case, the yield point occurred before
the buckling of the cylinder. This information indicates that the data prior to the yield
are valid for determining the compressive properties of each of the designs. Results
are presented in Table 4.1.

The modulus of elasticity and yield strength were averaged when there was more
than one set of data, and the % of solid columns indicate how the modulus of elasticity and yield strength correlate to that of the solid cylinder. Again, the buckling
load value in Table 4.2 is the value determined from correlating digital imagery with
stress results obtained from the MTS machine.

TABLE 4.1
Uniaxial Compression Test of Cylinders

Design	Modulus of Elasticity (GPa)	Modulus % of Solid	0.2% Yield (MPa)	Yield % of Solid	Buckling Load (MPa)
Solid	36.58	-	−598.52	-	−604.64
Primitive	3.13	8.55	−38.79	6.48	−50.22
Diamond	3.07	8.39	−36.08	6.03	−46.68
I-WP	5.64	15.42	−75.54	12.62	−88.77

TABLE 4.2
Uniaxial Compression Test of Primitive Cubes

Specimen	Relative Density (%)	Modulus of Elasticity (MPa)	0.2% Yield (MPa)	Plateau Stress (MPa)	Toughness (J/m³)
1	13.45	1,078.46	−21.57	−19.87	8.90
2	17.20	1,822.37	−36.42	−31.84	14.57
3	20.89	2,578.73	−51.50	−49.54	22.45

4.2 UNIAXIAL COMPRESSION OF CUBES

The second set of uniaxial compression testing was again accomplished on the AFIT UTM as described in Section 3.2.2.1, but due to the buckling seen in the cylinder testing, ASTM E9-19 was no longer followed. The geometry of the specimens was changed to cubes, the height-to-width ratio was changed to 1:1, and the testing process solely followed the procedure outlined in Section 3.2.2.2. A similar setup to that used in the first set of uniaxial compression tests was used to gather high-resolution photography and videography data, only the placement of the cameras was changed. Testing aimed to determine the modulus of elasticity, yield strength, yield point, and plateau strength. All four of the TPMS designs were fabricated for this stage of testing.

Once again, before laboratory testing, the actual relative density was compared with the predicted design relative density. The same procedures for determining the designed relative density and actual relative density used for the uniaxial compression testing of the cylinder specimens were used for this round of testing. As shown in Figure 4.6, the actual relative density for each of the designs was again higher than predicted but tended toward the expected value with an increase in surface thickness. The largest difference for the primitive designs was 1.6% denser than the computational prediction, with an average difference of 0.8% denser. The largest difference

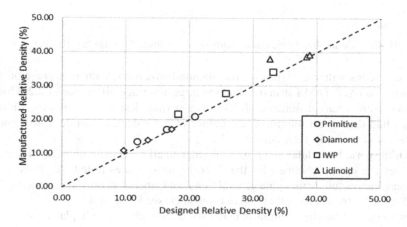

FIGURE 4.6 Comparison of cube design relative density.

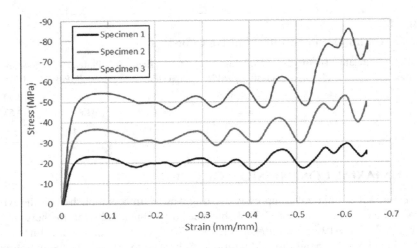

FIGURE 4.7   Uniaxial compression of primitive cell cubes.

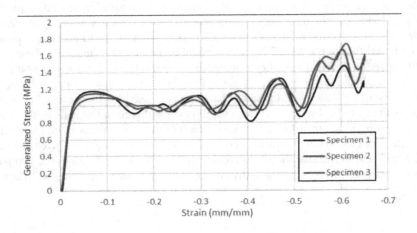

FIGURE 4.8   Generalized uniaxial compression response of primitive cell cubes.

was 1.2% denser than designed for the diamond structures, with an average of 0.5% denser. The I-WP TPMS structure had the largest average difference of 2.2% higher density, with a peak difference of 3.4% denser than designed. Lastly, the Lidinoid design had the largest maximum difference of 5.3% denser than designed, but the average was still only 1.9% denser.

All the periodic cellular cubes were tested, and all provided usable data. Figure 4.7 provides the results obtained for the three primitive cubes tested. All three of the response curves follow the same general trend, and when generalized against the 0.2% yield strength, the curves fall on top of each other; see Figure 4.8.

From these data, the modulus of elasticity, 0.2% yield strength, plateau stress, and toughness were determined. The methods used to determine these properties are those outlined in Section 3.2.2.3. The results are presented in Table 4.2.

TABLE 4.3

Poisson's Ratio for Primitive Cubes

Specimen	Poisson's Ratio
1	0.168
2	0.189
3	0.238

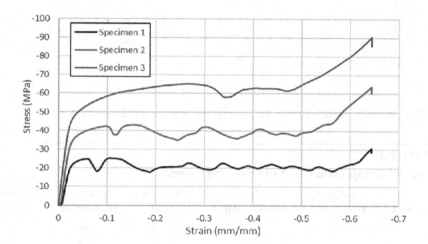

FIGURE 4.9 Uniaxial compression of diamond cell cubes.

Through digital image correlation, Poisson's ratio was also determined, and those results are presented in Table 4.3. Poisson's ratios seen for these designs are considerably less than for pure Inconel 718 (IN718), which is 0.30. Due to how the primitive surfaces collapse under a uniaxial load, there is less transverse strain.

Figure 4.9 provides the results obtained for the three diamond cubes tested. As with the primitive structures, the three response curves followed the same general trend, and when generalized against the 0.2% yield strength, the curves align rather well; see Figure 4.10.

Using these data, the modulus of elasticity, 0.2% yield strength, plateau stress, and toughness were determined. The methods used to determine these properties are those outlined in Section 3.2.2.3. The results are presented in Table 4.4.

Poisson's ratio was also determined using digital image correlation, and those results are presented in Table 4.5. As with the primitive periodic design, the diamond design saw substantially lower Poisson's ratios than the pure IN718 for all three specimens.

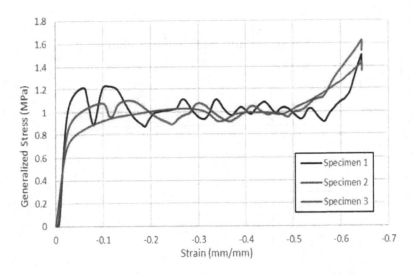

FIGURE 4.10   Generalized uniaxial compression response of diamond cell cubes.

TABLE 4.4
Uniaxial Compression Text of Diamond Cubes

Specimen	Relative Density (%)	Modulus of Elasticity (MPa)	0.2% Yield (MPa)	Plateau Stress (MPa)	Toughness (J/m³)
1	10.89	1,320.64	−26.43	−20.40	9.26
2	13.92	1,901.77	−37.95	−39.24	17.06
3	17.11	2,244.16	−44.91	−63.17	26.43

TABLE 4.5
Poisson's Ratio for Diamond Cubes

Specimen	Poisson's Ratio
1	0.134
2	0.171
3	0.227

Figure 4.11 provides the results obtained for the three I-WP cubes tested. As with the previous two periodic designs, the three response curves followed the same general trend. For the I-WP design, when generalized against the 0.2% yield strength, the curves align rather well through the plateau stress but then diverge once in the densification region; see Figure 4.12.

Using these data, the modulus of elasticity, 0.2% yield strength, plateau stress, and toughness were determined. The methods used to determine these properties are those outlined in Section 3.2.2.3. The results are presented in Table 4.6.

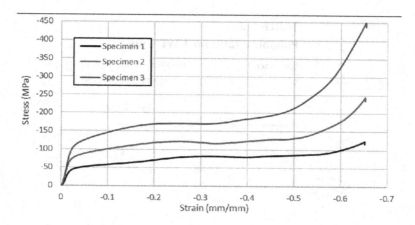

FIGURE 4.11 Uniaxial compression of I-WP cell cubes.

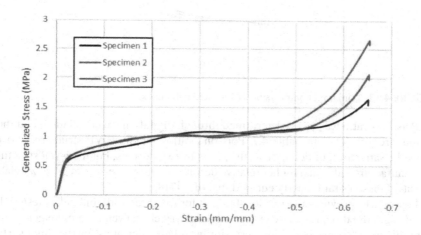

FIGURE 4.12 Generalized uniaxial compression response of I-WP cell cubes.

TABLE 4.6
Uniaxial Compression Test of I-WP Cubes

Specimen	Relative Density (%)	Modulus of Elasticity (MPa)	0.2% Yield (MPa)	Plateau Stress (MPa)	Toughness (J/m³)
1	21.62	2,558.26	−51.04	−75.49	30.70
2	27.79	4,067.09	−81.42	−118.37	49.04
3	34.27	5,515.41	−110.09	−169.56	70.91

TABLE 4.7

Poisson's Ratio for I-WP Cubes

Specimen	Poisson's Ratio
1	0.145
2	0.193
3	0.282

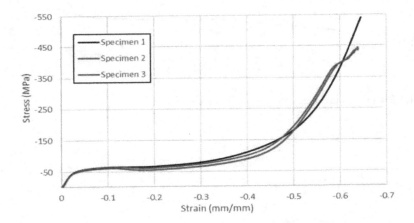

FIGURE 4.13   Uniaxial compression of Lidinoid cell cubes.

Poisson's ratio was again determined through digital image correlation, and those results are presented in Table 4.7. Poisson's ratios for the I-WP structures did not show the same level of decrease as the previous two designs, but the I-WP structures as manufactured also had higher relative densities. For this design, at 34.27% relative density, Poisson's ratio nearly equaled that of IN718.

Figure 4.13 provides the results obtained for the three Lidinoid cubes tested. For this design, the three response curves aligned very close even before the generalization; this was because of having very similar relative densities for the three surface thicknesses; see Table 4.9. For the Lidinoid design, when generalized against the 0.2% yield strength, the curves were nearly identical through the plateau stress, with relatively minor differences in the densification region; see Figure 4.14.

Using these data, the modulus of elasticity, 0.2% yield strength, plateau stress, and toughness were determined. The methods used to determine these properties are outlined in Section 3.2.2.3. The results are presented in Table 4.8.

Poisson's ratio was determined using digital image correlation, and those results are presented in Table 4.9. All three of the Lidinoid specimens had high relative densities, greater than 35%, and all three specimens had a greater Poisson's ratio than IN718. All three specimens displayed multiple different failure modes during compression, with significant fracturing evident in the video, and some crumbling was noted as well. These were likely the cause of Poisson's ratios exceeding that of pure IN718.

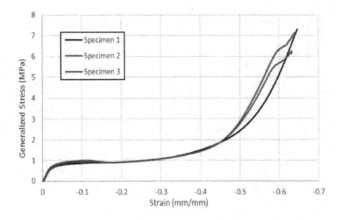

FIGURE 4.14 Generalized uniaxial compression response of Lidinoid cell cubes.

TABLE 4.8
Uniaxial Compression Test of Lidinoid Cubes

Specimen	Relative Density (%)	Modulus of Elasticity (MPa)	0.2% Yield (MPa)	Plateau Stress (MPa)	Toughness (J/m³)
1	37.95	2,455.68	−49.11	−73.96	34.35
2	38.65	2,499.45	−50.05	−62.24	29.54
3	39.23	2,689.46	−53.75	−69.40	32.48

TABLE 4.9
Poisson's Ratio for Lidinoid Cubes

Specimen	Poisson's Ratio
1	0.381
2	0.409
3	0.475

A comparison of the four designs was also accomplished through the use of specific material properties. The calculated material properties were compared with the properties of the base material, IN718, and then that ratio was plotted against relative density. Three material properties were compared, specific modulus of elasticity, specific plateau stress, and specific toughness. There were two common trends within the three comparisons. First, there is a need for more data to be collected over a more extensive relative density range for the primitive, diamond, and I-WP cellular designs. Second, the Lidinoid cellular structures did not perform at a comparable level with the other three designs, and therefore they will not be considered in further testing. The results for the I-WP structure align well with previously accomplished

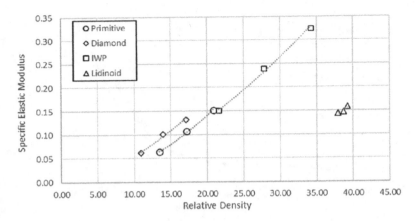

FIGURE 4.15   Specific elastic modulus.

work by Al-Rub et al. [106]. The other three structures have not been as widely researched, with no corresponding data available for comparison.

The specific modulus of elasticity was computed as the specimen modulus of elasticity divided by the modulus of elasticity of IN718. The comparison is shown in Figure 4.15. At first glance, it appears that the diamond structure performs the best of the designs, but as mentioned, there are no data over the same range as the I-WP design, and the primitive design appears to be trending with a steeper slope than the diamond. This could indicate that a diamond design may not perform as well as the others at a higher relative density.

The specific plateau stress was computed as the specimen plateau stress divided by the yield strength of the baseline material. The comparison across the four designs is shown in Figure 4.16. Again, across the lower relative densities, the diamond design provided the best performance. Still, there is no direct comparison of the diamond or primitive design at the relative densities of the test I-WP specimens or vice versa.

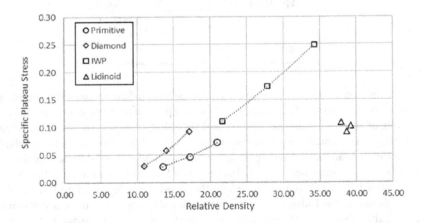

FIGURE 4.16   Specific plateau stress.

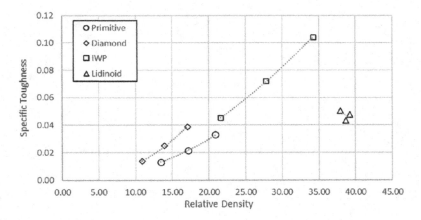

FIGURE 4.17 Specific toughness.

The specific toughness was computed as the specimen toughness divided by the yield strength of the baseline material. The comparison is shown in Figure 4.17. The relationships between the designs shown here are very similar to what was established in the results of the specific plateau stress. Again, more data must be collected across a shared relative density range to verify that the trends shown in these charts are maintained through the range of interest.

REFERENCES

88. ASTM International. Standard Test Methods of Compression Testing of Metallic Materials at Room Temperature, 2019.
105. A. Saada. *Elasticity: Theory and Applications*. J. Ross Publishing, Fort Lauderdale, FL, 2 editions, 1974.
106. R. K. Abu Al-Rub, D.-W. Lee, K. A. Khan, and A. N. Palazotto. Effective anisotropic elastic and plastic yield properties of periodic foams derived from triply periodic Schoen's I-WP minimal surface. *Journal of Engineering Mechanics*, 146(5): 04020030, 2020.

5 Mechanical Properties of Lattices and Design Variations

Based on the initial results from the testing described in Chapter 4, further investigation into the base lattice cells was performed. Additionally, derivative designs found through manipulation of the lattice cell trigonometric expressions were developed and tested. This research effort characterized both the base cell and variational design mechanical performance and deformation behavior, along with the comparison of the results across the source design.

5.1 MICROSTRUCTURAL ASSESSMENT

Following fabrication, the specimens were measured and weighed, and the actual relative densities were determined. Each of the triply periodic minimal surface (TPMS) designs and their variations were plotted to compare the actual relative density against the designed relative density. The designed relative density range for each specimen was determined by the use of three common surface thicknesses. This technique led to different relative density ranges for each specimen but was used to ensure consistent fabrication quality across the designs. A comparison between the as-designed relative density and the as-fabricated relative density can be found in Figure 5.1, with the diamond design variations depicted in Figure 5.1a, the I-WP design variations in Figure 5.1b, and the primitive design variations in Figure 5.1c. For all the designs, there was a greater difference between the actual and designed relative densities at lower density values [107].

This difference can be attributed to several factors, namely the intricacy of the design, variation of surface thickness, and bonding of the loose powder into the melt surfaces.

The effects of the design intricacy can be seen when comparing the trends of the diamond and primitive designs against the results of the I-WP design. The I-WP design has closer surface features, which are regions that are likely to have a greater surface thickness, as well as higher inadvertent fusing of loose powder. Some additional variation in surface thickness is due to the fabrication parameters, such as laser beam size and power. The laser parameters directly influence the size of the melt pool, which sets the fineness of the scan pattern and thereby the precision of the surface features. The melt pool parameters also have a direct impact on the amount of loose powder that will inadvertently bond to the structure.

Scanning electron microscopy (SEM) images were taken of manufactured parts for analysis of the print quality, which directly impact the actual relative density

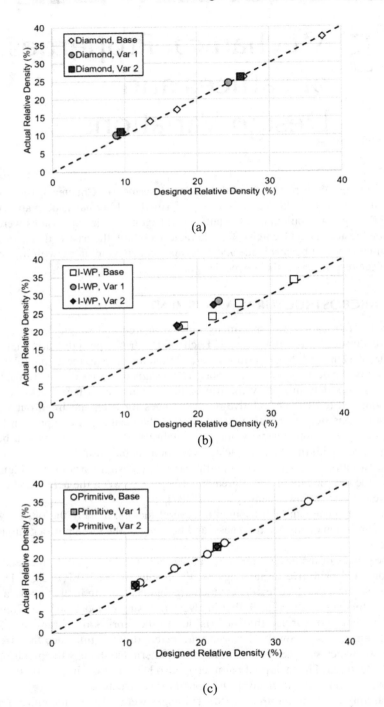

(a)

(b)

(c)

FIGURE 5.1   Designed versus actual relative density of (a) diamond design and variations, (b) I-WP design and variations, and (c) primitive design and variations.

of the specimens. A representative sampling of images for each of the three base cell designs is shown in Figure 5.2. As shown in these images, there is a significant amount of inadvertent fusing of powder to the cell surface. This additional material will increase the relative density of the printed part above that of the designed specimen. Another area of analysis is a comparison of the printed surface thickness to the designed surface thickness. For all three images, the designed surface thickness was 500 μm, and the SEM images were used with digital image correlation techniques to determine the actual surface thickness of the specimens. For the diamond cell design, Figure 5.2a, the actual surface thickness calculated was less than the designed value, with an average surface thickness of 476 μm. The combination of additional powder with a smaller surface thickness led to the actual relative density of the diamond cellular designs being remarkably close to the designed relative density. The I-WP cell design, Figure 5.2b, had an increased surface thickness compared

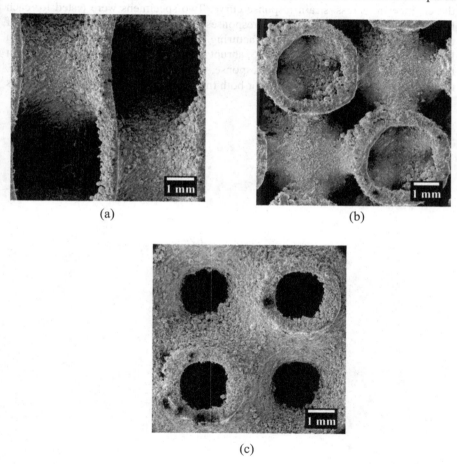

(a) (b)

(c)

FIGURE 5.2 Scanning electron microscope images showing the fabrication quality of the (a) diamond design, (b) I-WP design, and (c) primitive design.

with the designed value, having an average surface thickness of 524 µm. With the increased surface thickness and addition of partially fused powder, the I-WP actual relative density was greater than the designed relative density for all the manufactured specimens. Lastly, the primitive cell design, Figure 5.2c, also had an increased surface thickness compared with the designed thickness, although not as much as the I-WP, which had an average surface thickness of 0 507 µm. As with the I-WP design, the primitive design displayed an increased surface thickness and inadvertent powder fusing, which caused the actual relative density of the design to be greater than the designed relative density.

5.2 MECHANICAL PROPERTIES OF AS-BUILT LATTICES

The compressive mechanical properties of each of the lattices were derived from the engineering stress–strain response curve. Two specimens were tested for each lattice design, with the stress–strain responses being averaged to eliminate some of the variability due to additive manufacturing. The response curves are depicted in Figures 5.3–5.5. In Figures 5.3 and 5.5, abrupt changes in the stress–strain curve can be seen in the plateau region of the response, which is indicative of cellular failure. On the individual cell level, failure for both the diamond and primitive designs was

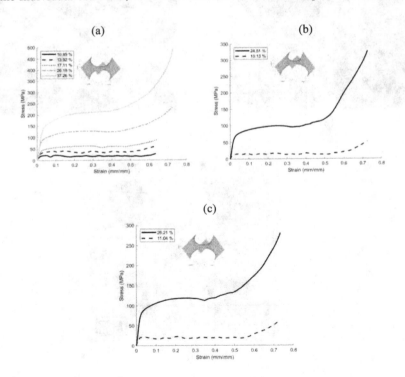

FIGURE 5.3 Uniaxial compression stress–strain response curves of diamond lattice designs: (a) base triply periodic minimal surface, (b) variation 1, and (c) variation 2.

typified by through-surface fracture, which can easily be seen in the 35% and 50% strain images in Figures 5.9–5.11 and 5.15–5.17. The surface fractures led to the stress drops in the plateau response, which was then followed by an increase in specimen stiffness that resulted in the subsequent rise in the stress response. In Figure 5.4, there is significantly less variation within the plateau region, which is consistent with the exhibited deformation behavior of the I-WP lattices. With the collective failure of cells being prevalent in the I-WP designs, individual cell failure was predominantly seen through bending and buckling of the lattice surface. All the stress–strain curves show a high level of consistency between the specific design's response. However, there are some noticeable differences between the different structural designs.

While the overall value of the plateau stress is predominantly dependent on the relative density of the design, the slope of the plateau region appears highly dependent on architecture. The three I-WP designs had the greatest plateau slope, which is likely due to early interaction between the cellular structures. The primitive designs displayed a nearly flat plateau region, indicating that there is little interaction between the cells before reaching densification strain. The diamond designs showed an intermediate response compared to the others.

The densification strain was highly dependent on the relative density of the specimen, in that densification occurred earlier in higher density specimens, but there

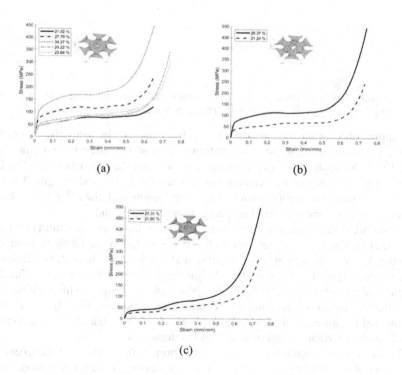

FIGURE 5.4 Uniaxial compression stress–strain response curves of I-WP lattice designs: (a) base triply periodic minimal surface, (b) variation 1, and (c) variation 2.

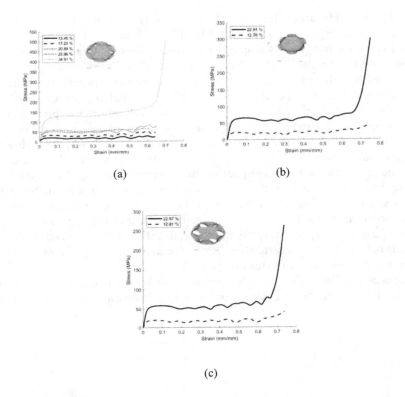

(a)   (b)

(c)

FIGURE 5.5 Uniaxial compression stress–strain response curves of primitive lattice designs: (a) base triply periodic minimal surface, (b) variation 1, and (c) variation 2.

were appreciable differences between the architectures again. The primitive cellular designs presented the highest densification strains of the three base architectures, again likely due to the lack of early interaction between the cell structures. The I-WP designs registered higher densification strains than the diamond designs. This may be due to the nature of deformation between the designs with the I-WP being through collective collapse and the diamond being through shear bands.

Additionally, there was a difference in the slope of the response within the densification region. The primitive responses also displayed the highest densification slope. With the delayed interaction between cellular structure, the rate of densification was considerably higher for the primitive design leading to the higher slope. The I-WP designs had the next highest densification slope of the designs. This may be due to the collective nature of deformation within these structures. Finally, the diamond designs had the lowest densification slope, which may be attributed to the combination of shear deformation and plateau region interaction.

The modulus of elasticity was determined through the analysis of the linear elastic portion of the response curve through a comparison of the tangent modulus and secant modulus, see Figure 5.6, as well as the more common least squares linear regression curve fit. The tangent modulus is determined as the slope of a line tangent

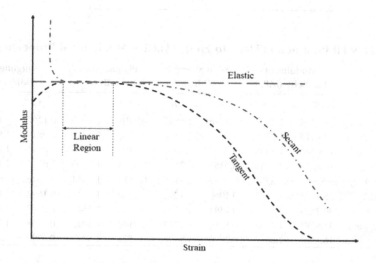

FIGURE 5.6 Method of determination for the elastic modulus from the tangent and secant moduli.

to the linear elastic stress–strain response, and the secant modulus is determined as the slope of a line from the origin that intersects the stress–strain response curve within the linear elastic region. The values obtained through both methods were within 1% of each other.

The yield strength was determined using 0.2% offset method. The plateau stress was calculated as the average stress value between 20% and 40% strain. This ensures that any peak stress, effects of densification, or inter-cell interactions that increase the compressive resistance of the structure are included in the plateau stress. The densification strain was determined through the analysis of the densification region of the stress–strain response, finding where the slope of the response intersects the abscissa, as shown in Figure 1.6. Finally, the toughness was obtained through numerical integration of the stress–strain response, using the trapezoid method, from load initiation to the densification strain. Toughness was chosen as a property of interest due to its relationship with the structure's ability to absorb energy. The mechanical properties were then plotted against the design relative density such that a power law relationship between the two values could be determined. Each mechanical property curve fit took the form of Equation 5.1. ϕ_{latt} is the mechanical property of the lattice structure under examination, ρ_{rel} is the lattice relative density, and C and n are fit coefficients.

$$\phi_{latt} = C\rho_{rel}^{n} \tag{5.1}$$

The summarized mechanical property power law fit coefficients for these lattices are presented in Table 5.1. Across the range of mechanical property curve fits, the average R^2 value was 0.9937.

TABLE 5.1
Power Law Fit Parameters Used to Fit the Lattice Mechanical Properties

Design	Modulus of Elasticity (MPa)		Yield Strength (MPa)		Plateau Stress (MPa)		Toughness (MJ/m³)	
	C	n	C	n	C	n	C	n
Diamond, base	79.1144	1.2143	0.4116	1.5841	0.2701	1.8700	0.1250	1.8749
Diamond, var. 1	47.9615	1.3706	0.2115	1.7566	0.1707	1.9918	0.0874	1.9799
Diamond, var. 2	108.4482	1.1338	0.3638	1.6266	0.1893	1.7960	0.0991	1.9581
I-WP, base	20.1234	1.6151	0.0355	2.2353	0.3004	1.7960	0.0377	2.1964
I-WP, var. 1	9.6266	1.8360	0.0987	1.9116	0.2087	1.8833	0.0646	2.0181
I-WP, var. 2	36.1133	1.2538	0.5966	1.1292	0.3742	1.5941	0.1694	1.6055
Primitive, base	30.2849	1.4497	0.1891	1.7174	0.1250	1.9545	0.0369	2.0985
Primitive, var 1	35.8583	1.4223	0.1856	1.7717	0.2144	1.8043	0.1158	1.7932
Primitive, var. 2	16.1382	1.6812	0.2700	1.5929	0.1521	1.8823	0.0792	1.8890

The exponential fit coefficient, n, for the modulus of elasticity provides some additional insight into the nature of the design's deformation behavior [108]. If the exponent value of the curve fit is approximately equal to one, then the lattice will, in general, carry the loading through tension or compression, which is considered stretching-dominated deformation. This indicates that the stiffness of stretch-dominated materials will change linearly with their relative density. However, if the exponent value is approximately equal to two, then the lattice will carry the loading through bending, buckling, or crushing, which is considered bending-dominated deformation. A common trend seen with bending-dominated deformation is that structural failure happens through the thickness of the surface. This means that the stiffness of bending-dominated materials will change quadratically with a change in their relative density. In general, the nature of loading and failure of the stretching-dominated lattices suggest that these architectures would perform better under uniaxial loading than the bending-dominated designs, yielding a higher strength-to-weight ratio. However, as mentioned, the plateau region of the bending-dominated structure response tended to be elongated compared with that of an equivalent stretching-dominated design, signifying that bending-dominated designs would perform better within energy-absorbing applications [5]. With these exponent values being between the stretching-dominated and bending-dominated values, there is a mixed mode of deformation, which was seen in Figures 5.9–5.17.

For comparison of the lattice designs, the mechanical properties of interest were plotted against the actual relative density for each specimen, shown in Figures 5.7 and 5.8. An overall trend seen across the four plots is that there is a greater effect of cellular design at lower relative densities, noted by the larger spread of points between designs. This is likely due to the underlying material properties becoming dominant within the mechanical response as the relative density increases.

For the elastic modulus, Figure 5.7a, the base diamond TPMS and its variations showed the best performance across the range of evaluated relative densities. At the upper end of the relative density range, the base I-WP and its first variation indicated similar results to the diamond, but again the tendency is for all of the results to align

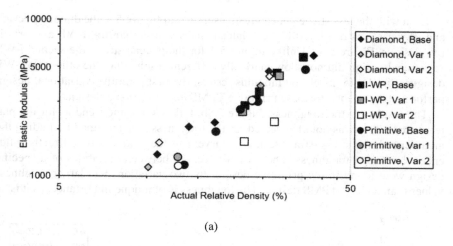

(a)

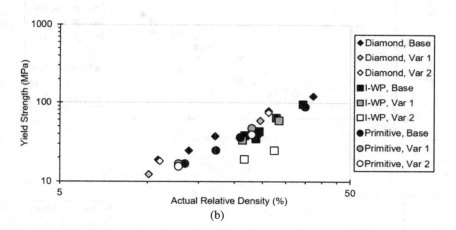

(b)

FIGURE 5.7 Comparison of experimental mechanical properties versus relative density of lattice designs: (a) elastic modulus and (b) yield strength.

with increasing relative density. All three of the primitive designs performed relatively consistent within the range as well, with the first primitive variation slightly outperforming the other two designs. The second variation of the I-WP TPMS exhibited significantly lower performance than the other lattice designs for elastic modulus, which may be attributed to difficulty manufacturing fine geometric features arising in this variational design.

Similar results were seen for the yield strength, Figure 5.7b. All three of the diamond designs performed the best over the entire range, with nearly identical results. The I-WP, its first variation, and the three primitive designs performed relatively equal to each other. Again, I-WP variation 2 exhibited considerably worse performance than the other eight lattices. This is consistent with the results from the elastic modulus since the modulus is used in determining the yield strength with the 0.2% offset method.

Even with the less stable plateau response, seen in Figure 5.3, the diamond design and its variations displayed higher plateau stress values than the I-WP and primitive designs, Figure 5.8a. Unlike Figure 5.7, for the plateau stress, the second I-WP variational design did not show markedly different results than the other two I-WP designs. Similar to the elastic modulus results, the first primitive variational design performed better than the base primitive TPMS and variation 2 design.

The results for the toughness, Figure 5.8b, follow the same trend as for the plateau stress. This was not unexpected, as the toughness is determined by finding the area beneath the stress–strain response curve, and the plateau stress directly influences the design toughness. The only differences between variations of a specific design were found in the primitive design. Its two variations had larger toughness values than its base TPMS design. The toughness is of unique importance, as it is an

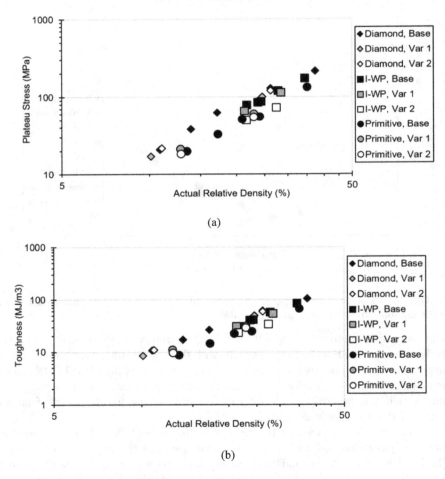

(a)

(b)

FIGURE 5.8 Comparison of experimental mechanical properties versus relative density of lattice designs: (a) plateau stress and (b) toughness.

indication of the ability of the structure to absorb energy, which can be balanced with a more productive strength-to-weight ratio.

5.3 DEFORMATION BEHAVIOR

Images representative of the structure responses under uniaxial compression are presented in Figures 5.9–5.17. The strain values depicted in these figures range from load initiation, 0% strain, to shortly before reaching the densification strain, 50% strain. This allows for analysis through the mechanical failure region of the lattice structure and not the material. It is worth noting that the relative density between the different lattices is not consistent. However, the failure patterns observed were consistent across relative densities for each of the lattice types. This indicates that the deformation behavior of lattice structures is remarkably independent of relative density but is decidedly dependent on the topology of the lattice. While the relative density of the structure does not play a significant role in the deformation behavior, it will impact the degree to which failure artifacts are present in the stress–strain response. At higher relative densities, the stress fluctuations that are indicative of cell failure will be less pronounced or even masked; see Figure 5.3a.

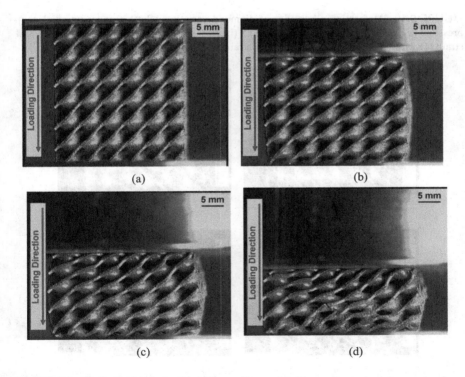

FIGURE 5.9 Deformation behavior of base diamond lattice under uniaxial compression: (a) 0% strain (0.0 mm displacement), (b) 20% strain (6.4 mm displacement), (c) 35% strain (11.2 mm displacement), and (d) 50% strain (15.9 mm displacement).

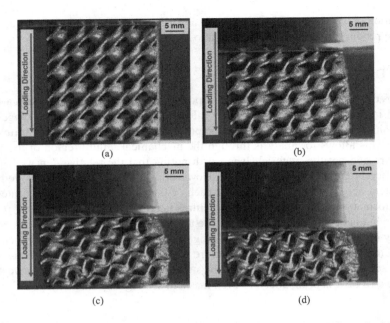

FIGURE 5.10 Deformation behavior of diamond variation 1 lattice under uniaxial compression: (a) 0% strain (0.0 mm displacement), (b) 20% strain (6.4 mm displacement), (c) 35% strain (11.2 mm displacement), and (d) 50% strain (15.9 mm displacement).

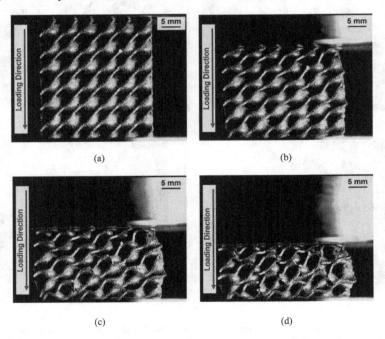

FIGURE 5.11 Deformation behavior of diamond variation 2 lattice under uniaxial compression: (a) 0% strain (0.0 mm displacement), (b) 20% strain (6.4 mm displacement), (c) 35% strain (11.2 mm displacement), and (d) 50% strain (15.9 mm displacement).

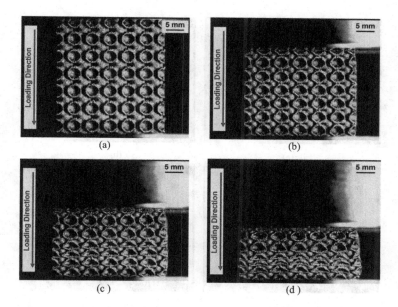

FIGURE 5.12 Deformation behavior of base I-WP lattice under uniaxial compression: (a) 0% Strain (0.0 mm displacement), (b) 20% strain (6.4 mm displacement), (c) 35% strain (11.2 mm displacement), and (d) 50% strain (15.9 mm displacement).

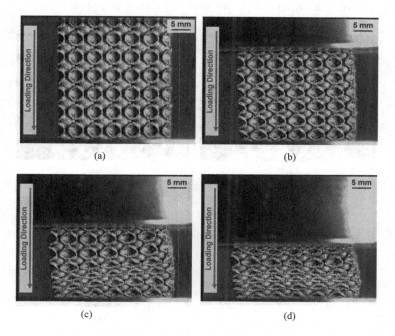

FIGURE 5.13 Deformation behavior of I-WP variation 1 lattice under uniaxial compression: (a) 0% strain (0.0 mm displacement), (b) 20% strain (6.4 mm displacement), (c) 35% strain (11.2 mm displacement), and (d) 50% strain (15.9 mm displacement).

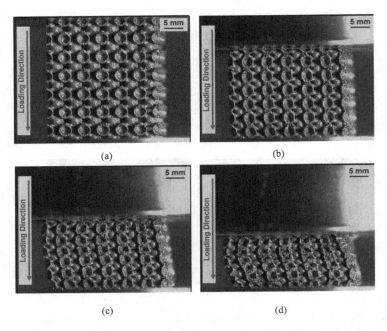

FIGURE 5.14 Deformation behavior of I-WP variation 2 lattice under uniaxial compression: (a) 0% strain (0.0 mm displacement), (b) 20% strain (6.4 mm displacement), (c) 35% strain (11.2 mm displacement), and (d) 50% strain (15.9 mm displacement).

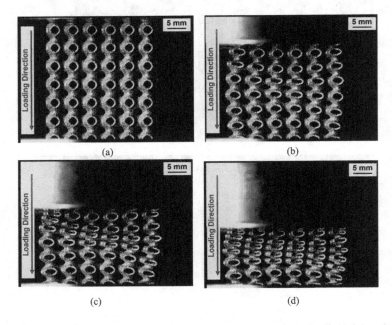

FIGURE 5.15 Deformation behavior of base primitive lattice under uniaxial compression: (a) 0% strain (0.0 mm displacement), (b) 20% strain (6.4 mm displacement), (c) 35% strain (11.2 mm displacement), and (d) 50% strain (15.9 mm displacement).

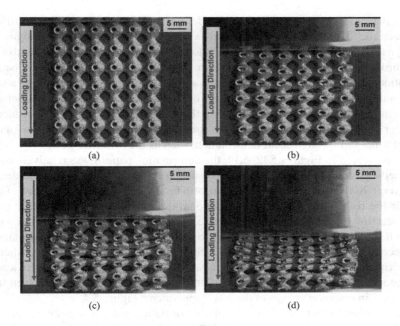

FIGURE 5.16 Deformation behavior of primitive variation 1 lattice under uniaxial compression: (a) 0% strain (0.0 mm displacement), (b) 20% strain (6.4 mm displacement), (c) 35% strain (11.2 mm displacement), and (d) 50% strain (15.9 mm displacement).

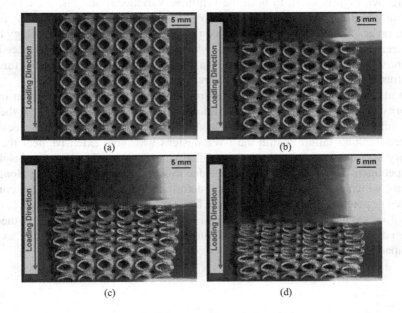

FIGURE 5.17 Deformation behavior of primitive variation 2 lattice under uniaxial compression: (a) 0% strain (0.0 mm displacement), (b) 20% strain (6.4 mm displacement), (c) 35% strain (11.2 mm displacement), and (d) 50% strain (15.9 mm displacement).

As depicted in Figures 5.9–5.11, the deformation pattern was slightly different between the diamond TPMS lattice and its variations. While all three designs indicated a shear failure pattern, the base diamond lattice showed dual shear bands forming throughout the strain range, or failure along two diagonal lines of cells. Both variational designs showed failure along a single shear band or failure along a single diagonal line of cells. Due to the topology of the diamond cell, single shear deformation can lead to cell layer collapse along that diagonal axis, which will reduce the number of stress fluctuations within the stress–strain response compared with other designs that fail through shear deformation, such as the primitive design.

As presented in Figures 5.12–5.14, the deformation pattern was relatively consistent between the I-WP TPMS lattice and its variations. In all three lattice designs, the overarching failure mechanism is uniform failure across a horizontal row of cells.

For the base I-WP and its first variational design, there is a near-failure of full rows starting at 35% strain; see Figures 5.12c and 5.13c. This deformation pattern holds through 50% strain. However, up to 20% strain for all three designs, and for the second variational design, there is a collective structure deformation, where all the horizontal rows are deforming in the same manner. This uniform failure leads to the smooth hardening-like response, with minimal stress fluctuations, as seen in the stress–strain response curves of the I-WP specimens, Figure 5.4.

As detailed in Figures 5.15–5.17, the deformation pattern was again relatively consistent between the base primitive TPMS lattice and its variations. For all three lattice designs, dual shear bands can be seen forming as early as 20% strain. For the base primitive lattice, one of the shear bands was more dominant than the other, see Figure 5.15c; however, as the strain accumulated, it maintained its dual shear failure; see Figure 5.15d. In both the primitive variational designs, the dual shear band failure progressed comparatively along both failure lines throughout the tested strain range. The shear deformation in conjunction with the primitive topology leads to increased stress concentration points, which in turn leads to more pronounced cell failure. These failure points can be seen in the stress–strain response curves through the stress fluctuations within the plateau region; see Figure 5.5.

In all cases, surface fracture began around 30% strain and can be seen in the majority of the 35% strain images. Additionally, it is interesting to note that the cellular failure is more prevalent near the loading surface in all three primitive designs, whereas cellular failure within the I-WP designs was more extensive near the stationary surface. The diamond designs failed nearly uniformly along the height of the specimen. These differences may be due to surface boundary loading along the platens, which could lead to constrained deformation due to friction on the contact surface or potential inconsistencies in the fabrication process that could lead to early failure in a region. Additionally, the cell topology likely plays a role, as the cell design dictates the amount of structural surface contact at the loading platen, as well as impacting the print quality.

REFERENCES

5. M. Scheffler and P. Colombo, editors. *Cellular Ceramics.* Wiley-VCH, Hoboken, New Jersey, 2005.

107. D. G. Spear, A. N. Palazotto, and R. A. Kemnitz. Mechanical properties of additively manufactured periodic cellular structures and design variations. *Journal of Engineering Materials and Technology*, 143(4), 05 2021. 041004.

108. V. S. Deshpande, M. F. Ashby, and N. A. Fleck. Foam topology: bending versus stretching dominated architectures. *Acta Materialia*, 49(6), 1035–1040, 2001.

REFERENCES

6 Split Hopkinson Pressure Bar Test Results

Split Hopkinson Pressure Bar (SHPB) testing was used to acquire mid-range strain rate data to be used in conjunction with the Taylor impact test data, as well as for initial insight into the utility of the Johnson–Cook (JC) Plasticity model. This required determining a baseline quasi-static material response with test specimens of the same size as used in the SHPB setup.

6.1 QUASI-STATIC MECHANICAL PROPERTIES OF TRIPLY PERIODIC MINIMAL SURFACE LATTICES

The quasi-static uniaxial compression testing was accomplished with seven specimens targeting four strain rates between 0.001 and 0.100 $\sec^{-1}$ for each of the lattice designs. The strain rates were calculated as the rate of change in specimen height, based on the set compression rate that was verified through digital image correlation, divided by the original height of the specimen. Across the range of evaluated strain rates, the mechanical response of the lattices appears relatively strain rate insensitive, which can be seen in the strain rate sensitivity curve presented in Figure 6.1. The strain rate sensitivity curve is based on the methodology to determine the JC strain rate sensitivity parameter, C, but only considering the data obtained within the quasi-static strain rate regime. As such, in Figure 6.1, the vertical axis is the ratio of the dynamic-to-static stress, and the horizontal axis is the natural log of the strain rate. Only the I-WP lattice design indicates minor strain rate sensitivity within the quasi-static range of the three lattice designs. The diamond lattice had a strain rate sensitivity parameter of 0.0015, whereas the I-WP lattice parameter was −0.0079, and the primitive lattice parameter was − 0.0001. The relative strain rate insensitivity aligns with the results found by Deshpande and Fleck, which evaluated metal foams under varying compression strain rates [118].

A typical stress–strain response curve of a surface-based lattice under uniaxial compression is depicted in Figure 6.2. The response curve consists of three distinct phases. The first phase is the linear-elastic response of the lattice, from load initiation up to the elastic yield. Here the modulus of elasticity (E) and yield strength (σ_y) are determined. The second phase is the plateau region, which consists of a plastic-like response of cell deformation under a near-constant load, which implies that the load is being distributed throughout the structure while the stiffness of some cells increases and that of others decreases. In this stage, the plateau stress (σ_{pl}) is determined. The final phase is densification, which is noted by a sharp rise in stress starting at the densification strain (ε_d) until complete structural failure. In this phase, all the cells within the structure have collapsed, and the material response is similar to that of a solid material.

DOI: 10.1201/9781003299639-6

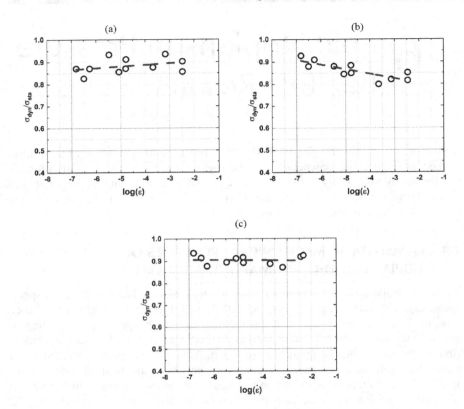

FIGURE 6.1   Quasi-static strain rate sensitivity of (a) diamond lattice, (b) I-WP lattice, and (c) primitive lattice.

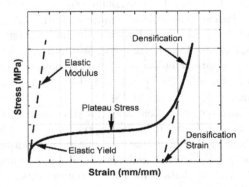

FIGURE 6.2   Typical stress–strain response curve of a surface-based lattice under uniaxial compression loading.

Within the quasi-static strain rate range, the diamond lattice design, Figure 6.3a, displays some variability in the response curves; however, the general trend of the responses holds across all of the specimens within ten percent for the stress values.

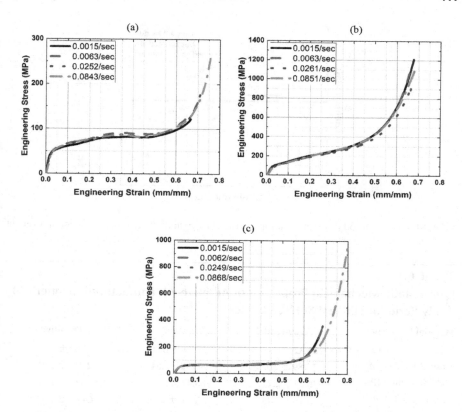

FIGURE 6.3 Quasi-static stress–strain response curves of (a) diamond lattice, (b) I-WP lattice, and (c) primitive lattice.

Similar to the diamond lattice, the I-WP lattice specimens show little variability within the quasi-static range. However, in the case of the I-WP design, the plateau region is never truly established due to the high relative density of these lattices. The response curves follow a more traditional elastic-plastic response until densification occurs. Again, the primitive lattice response curves, Figure 6.3c, show reduced variability compared with the diamond lattice curves, and they exhibit the typical foam/lattice response curve with well-defined elastic, plateau, and densification regions.

The modulus of elasticity, yield strength, plateau stress, and toughness were determined from the stress–strain response data for each specimen. The elastic modulus was determined through a comparison of the tangent modulus and secant modulus within the linear response region. Figure 6.4 provides an illustration of this process using an artificial data set. The tangent modulus is the slope of a line tangent to the stress–strain response within the linear-elastic region, and the secant modulus is the slope of a line from the origin of the stress–strain response that intersects the curve within the linear-elastic region. The yield strength was determined using the 0.2% offset method. Plateau stress was determined as the average stress value over the plateau regime of the stress–strain response. Finally, the toughness was calculated as the area beneath the response curve from the onset of loading up to the densification

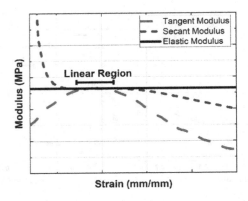

FIGURE 6.4 Method of determination for the elastic modulus from the tangent and secant moduli.

TABLE 6.1

Quasi-Static Mechanical Properties of Additively Manufactured Inconel 718 Triply Periodic Minimal Surface Lattices

Material Property	Diamond	I-WP	Primitive
Relative Density (%)	22.32 ± 0.19	40.66 ± 0.60	27.62 ± 0.37
Elastic Modulus (MPa)	2,789 ± 226	4,292 ± 664	2,475 ± 259
Yield Strength (MPa)	46.24 ± 2.72	83.47 ± 5.23	47.58 ± 2.51
Plateau Stress (MPa)	83.13 ± 2.48	238.06 ± 4.65	64.86 ± 1.80
Toughness (MJ/m³)	30.84 ± 0.74	79.86 ± 1.55	26.02 ± 0.73

strain. Toughness was chosen as a mechanical metric of interest as it can be viewed as a measure of how well the material, or in this case cellular design, can absorb or control energy. The results are provided in Table 6.1. As seen in this table, the relative density of the I-WP lattice is significantly higher than that of the other two designs, which is primarily the reason that the property values are so different.

6.2 DYNAMIC MECHANICAL PROPERTIES OF TRIPLY PERIODIC MINIMAL SURFACE LATTICES

The SHPB testing was accomplished targeting four strain rates, 400, 600, 800, and 1,000 sec⁻¹, with three specimens tested at each strain rate across each of the lattice designs. The data collected from the SHPB were used to calculate the true stress–strain response for each specimen; then the response was averaged across the common target strain rates. As mentioned, in evaluating the state of equilibrium of the specimen during testing, measurement of the forces acting on the specimen's front and back faces is required. Using the available data, the front and back face forces of the specimen were determined. Figure 6.5 presents the relationship between the force seen by the front face and the back face of the test specimen. As seen in this

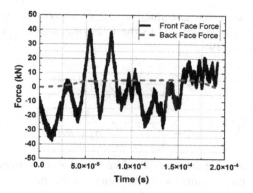

FIGURE 6.5 Representative front face force versus back face force for the diamond lattice at a target strain rate of 600 sec⁻¹.

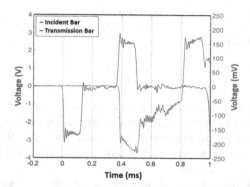

FIGURE 6.6 Representative strain gage signal for the diamond lattice at 1,000 sec⁻¹ strain rate

figure, there is a lot of noise present in the front face signal; this is predominantly due to an impedance mismatch between the incident bar and specimen.

The strain gauge signal data were recorded for each of the tests. Figure 6.6 provides a representative history of the diamond lattice design at a target strain rate of 1,000 sec⁻¹. As shown in Figure 6.6, the incident and reflected strain signals are of nearly equal magnitudes, which is a further indication of an impedance mismatch. With the reflected wave being so close in magnitude to the incident wave, it is difficult to measure the front face force of the specimen accurately.

Since it could not be determined that the front face force and back face force of the specimen were equal, the stress equilibrium relationship, $\sigma_I + \sigma_R = \sigma_T$, had to be used to determine the equilibrium of the specimen. This relationship can be rewritten as

$$\frac{\sigma_T}{\sigma_I + \sigma_R} \cong 1 \qquad (6.1)$$

Analysis of the incident bar specimen and transmission bar specimen interfaces using wave theory principles leads to the following stress relationships shown in Equations 6.2 and 6.3 [119]:

$$\sigma_T = \frac{2A_sZ_s}{A_sZ_s + A_bZ_b}\sigma_I = T\sigma_I \qquad (6.2)$$

$$\sigma_R = \frac{A_sZ_s - A_bZ_b}{A_sZ_s - A_bZ_b}\sigma_I = R\sigma_I \qquad (6.3)$$

In these equations, A is the cross-sectional area, and Z is the impedance, designated with subscript s for the specimen and subscript b for the incident or transmission bar. Impedance is calculated as the material wave velocity multiplied by the material density, $Z = C\rho$. T is designated the transmission coefficient, and R is the reflection coefficient. Substituting Equations 6.2 and 6.3 into Equation 6.1 provides the following:

$$\frac{T\sigma_I}{\sigma_I + R\sigma_I} \cong 1 \qquad (6.4)$$

which can be simplified to

$$\frac{T}{1+R} \cong 1 \qquad (6.5)$$

Using the material properties and geometries of the bars and specimen, the respective Z and A values were determined, and the expression was solved, with the result being $1.37 \cong 1$. Based on the previously mentioned works [92–94], it was concluded that the specimen was in a state of near-dynamic equilibrium and that Equations 3.24–3.26 were valid to be used.

With the state of dynamic equilibrium addressed, the specimen's strain and stress were determined. See Figure 6.7 for representative SHPB results. Figure 6.7a shows the true strain achieved versus time, and Figure 6.7b shows the true stress versus time for the specimen. As with the quasi-static range, the stress–strain response showed little variation across the dynamics strain rate range. However, there were some noticeable differences between the quasi-static and dynamic strain rates with the modulus of elasticity and yield strength.

In evaluating the dynamic strain rate data, the results obtained over the strain rate range tested indicated some effects from wave interactions. In addition to the expected waves present during SHPB testing, due to the curved surface–based network presented by the lattice structure on the interior of the specimen, there may be some wave reflection and loss within the specimen. Furthermore, under dynamic conditions, it has been found that multi-collapse modes can be activated simultaneously versus the singular and localized collapse modes typically observed in quasi-static testing [120]. These factors can be noted by some of the oscillations in the stress curve presented in Figure 6.7b.

Across the dynamic strain rate range tested, the diamond lattice, Figure 6.8a, showed a steady linear loading into the flow stress response; however, two of the

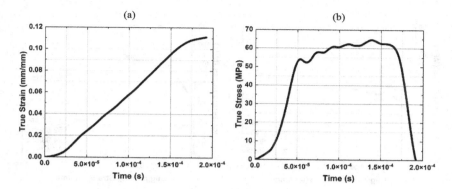

FIGURE 6.7   Split Hopkinson Pressure Bar results for the diamond lattice at a strain rate of 600 sec^{-1}: (a) true strain versus time, and (b) true stress versus time.

specimen responses indicated an earlier transition into the flow stress regime with regard to both stress and strain. Presenting a distinct regime transition followed by a more gradual stress rise indicates plastic deformation behavior. All the specimens exhibited a near-constant plateau region of the stress–strain response. The I-WP lattice dynamic response, Figure 6.8b, again showed a consistent linear loading region, presenting a more consistent transition strain than the diamond lattice. As with the quasi-static response curves, there does not appear to be a true plateau in the response, exhibiting what appears to be a strain hardening effect. However, the strain range achieved under the higher strain rates does not characterize the full response. The primitive lattice, Figure 6.8c, displayed the most consistent response across the four strain rates. As with the diamond lattice, the primitive lattice exhibited a very plastic deformation behavior. Similar to the quasi-static response, the primitive design had the lowest mechanical property values for estimated elastic modulus, flow stress transition, and plateau stress, but the values achieved were close to those of the diamond design.

While the SHPB data were primarily used to determine the plasticity model parameters, the mechanical properties were determined over the range of data acquired and are presented in Table 6.2. During SHPB testing, the specimens are not in equilibrium during the initial linear loading, which means that the data acquired in this regime will not provide an accurate analysis. However, as a point of comparison, the dynamic elastic modulus was estimated as the gradient of the stress–strain response curve as it approaches the zero point, Equation 6.6 [121]. There was a significantly broader spread in the dynamic elastic modulus than the quasi-static elastic modulus. This spread was primarily due to the specimen being in a state of non-equilibrium over the linear-elastic regime. However, this is also likely due in part to some relatively minor strain rate effects taking place over the larger spread of strain rates as a result of inertia effects and localized deformation.

$$E_d = \lim_{x \to 0}\left(\frac{\partial \sigma}{\partial \varepsilon}\right)_\sigma \qquad (6.6)$$

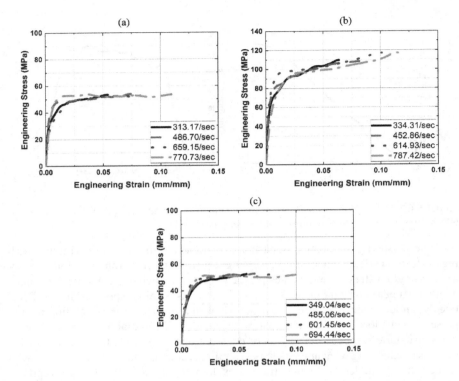

FIGURE 6.8 Dynamic stress–strain response curves of (a) diamond lattice, (b) I-WP lattice, and (c) primitive lattice.

TABLE 6.2

Dynamic Mechanical Properties of Additively Manufactured Inconel 718 Triply Periodic Minimal Surface Lattices

Material Property	Diamond	I-WP	Primitive
Relative Density (%)	22.32 ± 0.19	40.66 ± 0.60	27.62 ± 0.37
Elastic Modulus (MPa)	17,194 ± 6,449	34,222 ± 5,762	16,072 ± 5,660
Flow Stress (MPa)	41.22 ± 7.44	77.49 ± 6.45	39.08 ± 2.55
Plateau Stress (MPa)	52.42 ± 0.10	109.77 ± 3.49	51.20 ± 0.63

6.3   PLASTICITY MODEL PARAMETERS

Each of the three lattice designs was evaluated separately to determine the JC flow stress parameters. Table 6.3 presents the quasi-static flow stress parameters, and the full-model results can be found in Table 6.4. For each lattice, the reference strain rate chosen was 0.085 sec⁻¹. As mentioned, the parameter A for each design was determined as the quasi-static 0.2% yield strength value at the chosen reference strain. Parameters B and n were also determined at the reference strain value, so the

TABLE 6.3

Quasi-Static Johnson–Cook Flow Stress Parameters of Additively Manufactured Inconel 718 Triply Periodic Minimal Surface Lattices

Flow Stress Parameter	Diamond	I-WP	Primitive
A (MPa)	46.01	88.41	48.03
B (MPa)	169.65	1,186.65	60.98
n	0.7251	1.1006	0.3649
C	0.0015	−0.0079	−0.0001
m	−0.6169	0.1793	1.7829

TABLE 6.4

Johnson–Cook Flow Stress Parameters of Additively Manufactured Inconel 718 Triply Periodic Minimal Surface Lattices

Flow Stress Parameter	Diamond	I-WP	Primitive
A (MPa)	46.01	88.41	48.03
B (MPa)	169.65	1,186.65	60.98
n	0.7251	1.1006	0.3649
C	−0.0039	−0.0008	−0.0085
m	0.4928	0.4425	1.0441

quasi-static parameter representation and the full-model values were the same for the first three parameters. The values for C show that, except the I-WP lattice, the quasi-static strain rate sensitivity value is less than that of the full model. For the diamond lattice, the quasi-static C value indicated a strain rate hardening effect, which was the only lattice design to depict an increase in flow stress with an increase in strain rate. The values for m in the quasi-static range are based on a small ambient temperature range and at the slower compression speed have minor rate-temperature effects. The increased strain rates used in the determination of the full-model JC flow stress parameters include a greater range of temperature effects and data points, which increases the validity of these values compared with the quasi-static values.

Figure 6.9a depicts the plastic hardening response of the diamond lattice, and from this curve fit, the parameters B and n are determined. The linear regression for these data is $\log(\sigma_{pl}) = 0.7251\log(\varepsilon_{pl}) + 2.2296$, which results in $B = 169.65$ and $n = 0.7251$. Figure 6.9b shows the regression fit used to determine the C parameter for the diamond lattice. The slope of the regression is -0.0040, which is the C value. This indicates that the values for elastic modulus, yield strength or flow stress, plateau stress, and toughness of the diamond lattice will decrease at a higher strain rate. The diamond flow stress parameter, m, was found using the linear fit found in Figure 6.9c. Here the slope of the regression, m value, was calculated to be 0.4988.

The plastic response of the I-WP lattice design used in determining the B and n parameters is shown in Figure 6.10a. The linear regression fit for the I-WP data is $\log(\sigma_{pl}) = 1.1006\log(\varepsilon_{pl}) + 3.0487$, which results in $B = 1118.65$ and $n = 1.1006$. Figure

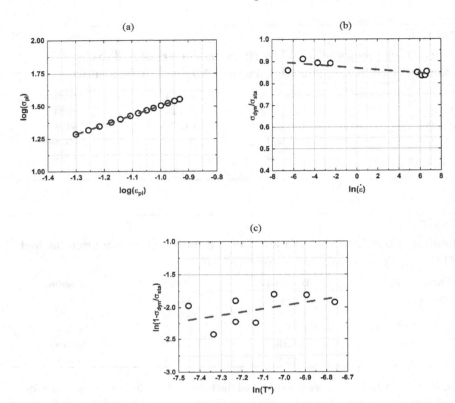

FIGURE 6.9 Johnson–Cook flow stress parameters for the diamond lattice: (a) B and n, (b) C, and (c) m.

6.10b shows the fit used to determine the flow stress C parameter for the I-WP lattice. The C value was determined to be -0.0008. As with the diamond lattice, the negative C value indicates that, like the diamond lattice, the elastic modulus, yield strength or flow stress, plateau stress, and toughness of the I-WP lattice will decrease at an increasing strain rate. The final flow stress parameter, m, for the I-WP lattice was found using the data presented in Figure 6.10c. Here the value for m was calculated to be 0.4604.

Figure 6.11a depicts the plastic regime used to calculate the B and n flow stress parameters for the primitive lattice. The expression for the linear regression fit is log $(\sigma_{pl}) = 0.3649\log(\varepsilon_{pl}) + 1.7852$, which results in $B = 60.98$ and $n = 0.3649$. The JC flow stress parameter C for the primitive lattice was determined from the data presented in Figure 6.11b. The C value was calculated to be -0.0085. Consistent with the other two lattice designs, the negative C value indicates that the elastic modulus, yield strength or flow stress, plateau stress, and toughness of the primitive lattice will decrease at an increasing strain rate. The final primitive lattice flow stress parameter, m, was found from the data depicted in Figure 6.11c. The m value for the primitive

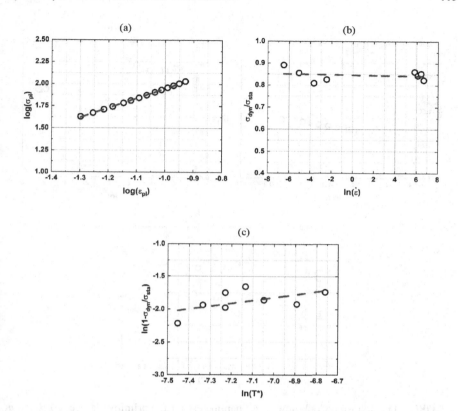

FIGURE 6.10 Johnson–Cook flow stress parameters for the I-WP lattice: (a) *B* and *n*, (b) *C*, and (c) *m*.

lattice was calculated to be 1.0438. The higher *m* value for the primitive lattice suggests that this lattice design is more sensitive to temperature than the diamond or I-WP lattices.

The negative strain rate effect seen for each of the three lattices is an interesting phenomenon. One potential source for this phenomenon is that the strain rate sensitivity determined here is not directly related to the applied strain rate, but rather at elevated strain rates, there is a tendency for localization of plastic flow near the stress concentration point located within the lattice [122]. The plastic work being done on the lattice during deformation leads to an increase in temperature around the concentration points, which could lead to localized thermal softening [72]. The bounded thermal softening will exacerbate the plastic flow near the stress concentration point, which ultimately leads to an earlier failure of the lattice structure at higher strain rates. Additionally, for other additively manufactured metal alloys, the makeup of the alloy powder has been tied to strain rate sensitivity through deformation resistance

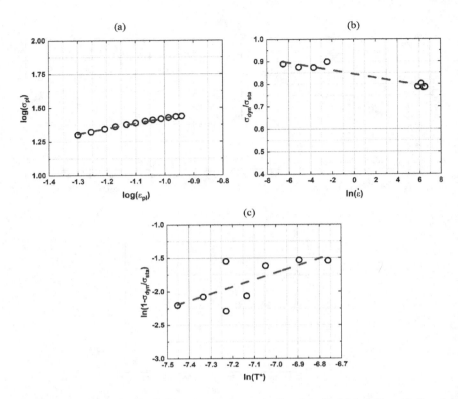

FIGURE 6.11   Johnson–Cook flow stress parameters for the primitive lattice: (a) B and n, (b) C, and (c) m.

brought about by pinning of dislocation. Pinning of dislocation is largely dependent on the diffusion rate of atoms that can be widely affected by both temperature and strain rate [123,124].

The resultant flow stress plasticity curves were compared against the experimental results as a measure of their goodness of fit. As the strain and stress ranges were varied for the quasi-static and dynamic responses, the strain rate regimes were split for the comparison charts. The quasi-static responses are depicted in Figure 6.12. As seen in these figures, the flow stress model closely matches the experimental true stress–strain response through the initial plastic hardening regime. However, the model falls apart at the onset of lattice densification as the model response was not developed to include a secondary stiffening. For low strains, up to approximately 30% strain, the model presents a good approximation of the lattice's viscoplastic response.

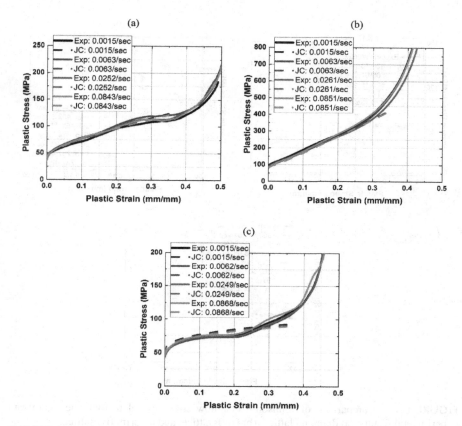

FIGURE 6.12 Comparison of Johnson–Cook flow stress model to quasi-static strain rate experimental results: (a) diamond lattice, (b) I-WP lattice, and (c) primitive lattice.

The dynamic strain rate responses are depicted in Figure 6.13. In all cases, the flow stress value tends to underestimate the experimentally observed values for the onset of plastic stress. Additionally, the model plastic hardening slope is steeper than the plateau stress rise observed in experimentation. Over the tested strain range, the model provides a proximate estimate of the lattice plastic, or plastic-like, response at the elevated strain rates, although errors up to 20% were observed. As densification does not occur for the dynamic response regime within the strain ranges achieved for these strain rates, there is no concern with modeling this phenomenon. However, as the impact velocities and strain rates increase, the response will progressively exhibit greater densification behavior.

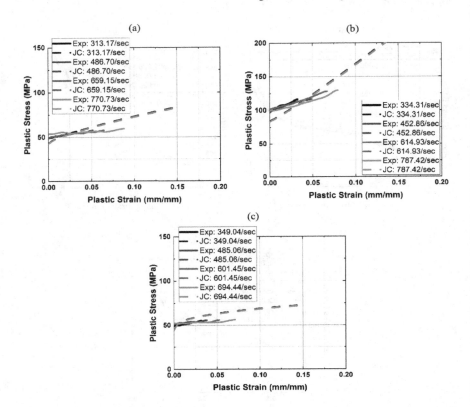

FIGURE 6.13 Comparison of Johnson–Cook flow stress model to dynamic strain rate experimental results: (a) diamond lattice, (b) I-WP lattice, and (c) primitive lattice.

REFERENCES

74. H. W. Meyer and D. S. Kleponis. Modeling the high strain rate behavior of titanium undergoing ballistic impact and penetration. *International Journal of Impact Engineering*, 26(1–10):509–521, 2001.

92. E. W. Billington and C. Brissenden. Mechanical properties of various polymeric solids tested in compression. *International Journal of Mechanical Sciences*, 13(6):531–545, 1971.

93. U.S. Lindholm. Some experiments with the split Hopkinson pressure bar. *Journal of the Mechanics and Physics of Solids*, 12(5):317–335, 1964.

94. U. S. Lindholm and L. M. Yeakley. High strain-rate testing: Tension and compression. *Experimental Mechanics*, 8:1–9, 1968.

118. V. S. Deshpande and N. A. Fleck. High strain rate compressive behaviour of aluminum alloy foams. *International Journal of Impact Engineering*, 24(3): 277–298, 2000.

119. W. Johnson. *Impact Strength of Materials*. Edward Arnold, London, United Kingdom, 1972.

120. M. Peroni, G. Solomos, and V. Pizzinato. Impact behaviour testing of aluminum foam. *International Journal of Impact Engineering*, 53: 74–83, 2013.

121. A. Massaq, A. Rusinek, and M. Klosak. Method for determination of the dynamic elastic modulus for composite materials. *Engineering Transactions*, 61: 301–315, 2013.
122. E. N. Borodin, A. A. Gruzdkov, A. E. Mayer, and N. S. Selyutina. Physical nature of strain rate sensitivity of metals and alloys at high strain rates. *Journal of Physics: Conference Series*, 991: 012012, 2018.
123. H. Yamada, T. Kami, R. Mori, T. Kudo, and M. Okada. Strain rate dependence of material strength in AA5xxx series aluminum alloys and evaluation of their constitutive equation. *Metals*, 8(8): 576, 2018.
124. P. Jakkula, G. Ganzenmu"ller, F. Gutmann, A. Pfaff, J. Mermagen, and S. Hiermaier. Strain rate sensitivity of the additive manufacturing material scalmalloy. *Journal of Dynamic Behavior of Materials*, 7(4): 518–525, 2021.

7 Strain Rate–Sensitive Constitutive Model for Lattice Structure

The quasi-static and dynamic compression testing showed that traditional constitutive models were not adequate for modeling and simulation of the lattice structures under high strain rates that led to large deformations. Therefore, a new constitutive model had to be developed for the research to move forward.

7.1 LEGACY MATERIAL MODELS

7.1.1 QUASI-STATIC RESPONSE MODELS

Rusch: The Rusch model consists of a relatively simple formulation consisting of a summation of two power laws [81,82]. The first bracketed term defines the response over the linear elastic and plateau regimes, and the second bracketed term defines the densification region of the response.

$$\sigma(\varepsilon) = \left[a\varepsilon^m \right] + \left[b\varepsilon^n \right] \tag{7.1}$$

Avalle et al. formulated their quasi-static model to improve on the Rusch model primarily in the elastic and plateau regions of the stress–strain response [84]. As with the Rusch model, the first bracketed term describes the elastic and plateau regimes, and the second bracketed term describes the densification response.

$$\sigma(\varepsilon) = \left[A\left\{ 1 - e^{-(E/A)\varepsilon(1-\varepsilon)^n} \right\} \right] + \left[B\left(\frac{\varepsilon}{1-\varepsilon} \right)^n \right] \tag{7.2}$$

Goga and Hučko developed a new phenomenological model derived from rheological models [125]. The linear elastic region was modeled using a serial Maxwell element, consisting of a linear spring and viscous damper. A second spring was added to the model in parallel with the original Maxwell element for the plateau region. Finally, the densification region of the response was modeled by incorporating a non-linear spring shown in Equation 7.3.

$$\sigma(\varepsilon) = e^{(-E_1/\eta_1)\varepsilon} \left(-1 + e^{(E_1/\eta_1)\varepsilon} \right) \eta_1 + \left[E_2 + \gamma \left(1 - e^\varepsilon \right)^h \right] \varepsilon \tag{7.3}$$

DOI: 10.1201/9781003299639-7

7.1.2   Dynamic Response Models

Johnson and Cook developed their flow stress model to describe the plastic response of metals under axial loading and incorporated strain rate and temperature effects [44]. The first bracketed term accounts for strain hardening or plastic strain accumulation. The second bracketed term accounts for the effects of strain rate. The final bracketed term accounts for the effects of temperature. Here T^* represents the homologous temperature, a ratio of the difference between the current temperature and room temperature over the difference between the material melting temperature and the current temperature shown in Equation 7.4.

$$\sigma(\varepsilon) = \left[A + B\varepsilon^n \right] \left[1 + C \ln\left(\frac{\dot{\varepsilon}}{\dot{\varepsilon}_0} \right) \right] \left[1 - \left(T^* \right)^m \right] \tag{7.4}$$

Jeong et al. developed a strain rate–sensitive model by incorporating a strain rate effect term into the Avalle quasi-static model [126].

$$\sigma(\varepsilon) = \left[A\left\{ 1 - e^{-(E/A)\varepsilon(1-\varepsilon)^n} \right\} + B\left(\frac{\varepsilon}{1-\varepsilon} \right)^n \right] \left[1 + (a + b\varepsilon) \ln\left(\frac{\dot{\varepsilon}}{\dot{\varepsilon}_0} \right) \right] \tag{7.5}$$

Avalle et al. expanded their previous work, proposing a new model incorporating strain rate effects [85].

$$\sigma(\varepsilon) = \left[\sigma_P \left(1 - e^{-m\varepsilon} \right) \right] + \left[\sigma_S \varepsilon \right] + \left[\sigma_D \varepsilon^n \right] \tag{7.6}$$

In this equation, $\sigma_P, \sigma_S,$ and σ_D are strain rate and density-sensitive expressions that describe the plateau stress, linear hardening slope, and densification of the response, respectively, indicated below.

$$\sigma_P = \sigma_{P,0} \left[1 + \left(\frac{\dot{\varepsilon}}{\dot{\varepsilon}_0} \right)^p \right] f_P(\rho) \tag{7.7}$$

$$\sigma_S = \sigma_{S,0} \left[1 + \left(\frac{\dot{\varepsilon}}{\dot{\varepsilon}_0} \right)^p \right] f_S(\rho) \tag{7.8}$$

$$\sigma_D = \sigma_{D,0} \left[1 + \left(\frac{\dot{\varepsilon}}{\dot{\varepsilon}_0} \right)^p \right] f_D(\rho) \tag{7.9}$$

7.2   PROPOSED FLOW STRESS MODEL

7.2.1   Model Development

The proposed model was developed to account for the high strain rate and temperature effects present in the high-speed impact of lattice structures. The model started from the Rusch equation, Equation 7.10, under quasi-static condition. It provides a good approximation for the stress–strain response using a simple formulation.

$$\sigma(\varepsilon) = A\varepsilon^m + B\varepsilon^n \tag{7.10}$$

The first modification made follows the early work of Avalle and Goga, whereas the strain in the second bracketed term is replaced by a strain ratio to reduce the model error in the densification transition [84,125] as shown in Equation 7.11.

$$\sigma(\varepsilon) = A\varepsilon^m + B\left(\frac{\varepsilon}{1-\varepsilon}\right)^n \tag{7.11}$$

Next, strain rate effects were applied to the formulation in a manner similar to that of Johnson and Cook.

$$\sigma(\varepsilon) = \left[A\varepsilon^m + B\left(\frac{\varepsilon}{1-\varepsilon}\right)^n \right]\left[1 + C\ln\left(\dot{\varepsilon}^*\right)\right] \tag{7.12}$$

where $\dot{\varepsilon}^*$ is the strain rate ratio of the applied strain rate to a reference quasi-static strain rate. Further analysis showed that each of the four variables present within the model varied differently with a change in strain rate, which led to the following representation.

$$\sigma(\varepsilon) = A\left(1 + C_1\ln\left(\dot{\varepsilon}^*\right)\right)\varepsilon^{m\left(1 + C_2\ln\left(\dot{\varepsilon}^*\right)\right)} + B\left(1 + C_3\ln\left(\dot{\varepsilon}^*\right)\right)\left(\frac{\varepsilon}{1-\varepsilon}\right)^{n\left(1 + C_4\ln\left(\dot{\varepsilon}^*\right)\right)} \tag{7.13}$$

Finally, temperature effects were added to the model using the same expression as the Johnson and Cook model.

$$\sigma(\varepsilon) = \left[A\left(1 + C_1\ln\left(\dot{\varepsilon}^*\right)\right)\varepsilon^{m\left(1 + C_2\ln\left(\dot{\varepsilon}^*\right)\right)} + B\left(1 + C_3\ln\left(\dot{\varepsilon}^*\right)\right)\left(\frac{\varepsilon}{1-\varepsilon}\right)^{n\left(1 + C_4\ln\left(\dot{\varepsilon}^*\right)\right)} \right]\left[1 - T_h^p\right] \tag{7.14}$$

Within this flow stress model, the terms A and m describe the yield point and plateau region of the stress–strain response, respectively, and B and n describe the densification transition curve and slope, respectively. As these four terms are all strain rate sensitive, the C_i terms are the strain rate sensitivity parameters. Finally, p is the temperature sensitivity parameter.

7.2.2 DETERMINATION OF MODEL PARAMETERS

The first step in determining the model parameters is to ascertain the quasi-static constants: A, m, B, and n. These parameters require the use of the true stress–strain response of the material, which is found using the relationships provided in Equations 7.15 and 7.16.

$$\varepsilon_t = \ln(1+\varepsilon_e) \tag{7.15}$$

$$\sigma_t = \sigma_e * (1 + \varepsilon_e) \tag{7.16}$$

The true stress–strain curve for the reference strain rate case, 0.010 sec⁻¹, of the diamond lattice is presented in Figure 7.1. Data from the diamond lattice will be used to detail the methodology; however, parameter results for the two remaining lattice designs will also be provided.

Parameters A and m are found by taking the natural log of true stress and true strain over the response region from the yield through the linear plateau region, Figure 7.2 for the diamond lattice, this was 0.03–0.30 strain. More than 400 data points from experimentation fell within this strain range to be used in defining the

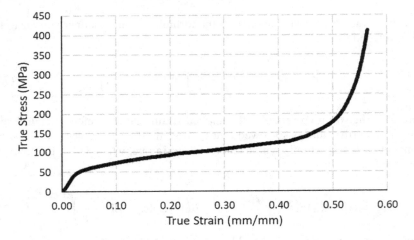

FIGURE 7.1 True stress versus true strain response of diamond lattice at 0.010 sec⁻¹ strain rate.

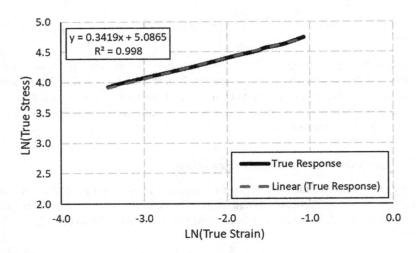

FIGURE 7.2 Determination of model parameters A and m.

curve of interest. This curve is then fitted with a linear regression line. m is found as the slope of the curve fit, and A is calculated as $e^{\text{intercept}}$. Here A is 161.82 and m is 0.3419.

Next, the plastic stress–strain response is calculated by subtracting the yield strain from the true strain (Equation 7.17) and subtracting the elastic stress from the true stress using A and m (Equation 7.18).

$$\varepsilon_{pl} = \varepsilon_t - \varepsilon_y \tag{7.17}$$

$$\sigma_{pl} = \sigma_t - A\varepsilon_t^m \tag{7.18}$$

Parameters B and n are then found by taking the natural log of the plastic stress and plastic strain over the linear region of the response above the densification strain, which was 0.55 strain for the diamond lattice; see Figure 7.3. Within the recorded data, more than 100 data points were available for use in fitting above the densification strain; 104 data points were used in the diamond curve. Again, this curve is fitted with a linear regression, where n is the regression slope and B is half of the value for $e^{\text{intercept}}$. For the diamond lattice, the parameters were found to be 6.7176 for n and 50.24 for B.

Once the quasi-static model parameters have been found, the strain rate sensitivity terms, $C_1–C_4$, are determined. C_1 is determined by plotting the dynamic to static stress ratio against the natural log of the strain rate ratio at a common point in the early plateau region, then finding the slope of a linear regression fit of the data. The dynamic stress is the experimental stress value, and the static stress is the model stress value using only the quasi-static parameters. The resultant plot for the diamond lattice is depicted in Figure 7.4. Here the value of C_1 was determined to be 0.0540.

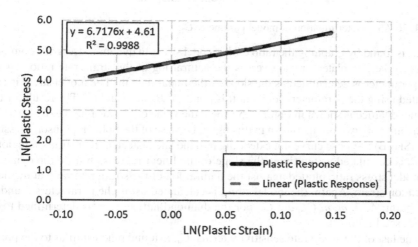

FIGURE 7.3   Determination of model parameters B and n.

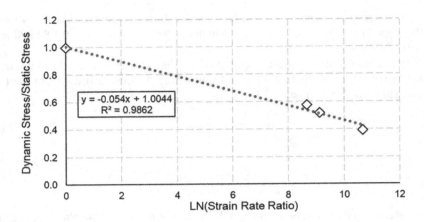

FIGURE 7.4 Determination of model parameter C_1.

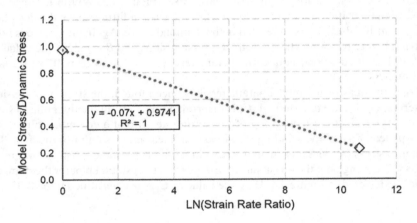

FIGURE 7.5 Determination of model parameter C_2.

C_2 is found by graphing the ratio of the model stress to dynamic stress at a common point in the late plateau region versus the natural log of the strain rate ratio. Again, the parameter is determined as the slope of the linear data fit. The model stress is calculated using the parameters found at this point: A, B, m, n, and C_1. The result for the diamond lattice is shown in Figure 7.5, where the value of C_2 was found to be -0.0700. There are only two points shown in this figure because of the Split Hopkinson Pressure Bar (SHPB) response curve not extending into the late plateau or densification regions.

C_3 is calculated as ten times the slope of the linear regression fit to the dynamic to model stress ratio plotted against the natural log of the strain rate ratio at the densification strain. Again, the model stress is calculated using the parameters found to this point: A, B, m, n, C_1, and C_2. For the diamond lattice, C_3 was determined to be 2.265, see Figure 7.6.

The last of the strain rate sensitivity terms, C_4, is found much similar to the previous three parameters. The model stress to dynamic stress ratio is again plotted versus

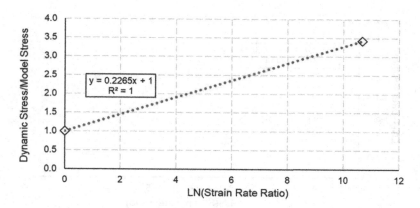

FIGURE 7.6 Determination of model parameter C_3.

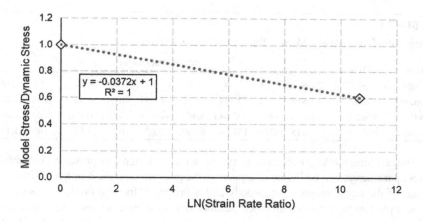

FIGURE 7.7 Determination of model parameter C_4.

the natural log of the strain rate ratio at a point within the plateau to densification transition. Then the data are fitted with a linear regression. The model stress now includes the C_3 term in its calculation. C_4 was determined to be −0.0372 for the diamond lattice; see Figure 7.7.

Finally, the temperature sensitivity parameter, p, is determined. p is calculated by plotting the natural log of the ratio of the thermal inclusive stress to room temperature stress versus the natural log of the homologous temperature. The thermal inclusive stress values are the experimental stress values, which include thermal effects, and the room temperature stress is calculated using calculated values of A, B, m, n, C_1, C_2, C_3, and C_4. Once again, the model parameter is determined as the slope coefficient of the linear regression fit, which was found to be 1.2853; see Figure 7.8. As with the temperature variable within the Johnson–Cook Flow Stress formulation, the temperature sensitivity parameter here is related to the plastic strain energy presumed within the lattice.

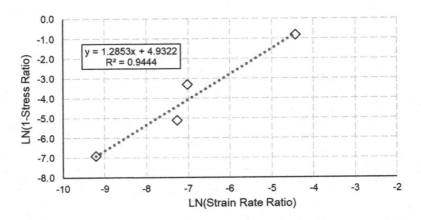

FIGURE 7.8 Determination of model parameter p.

TABLE 7.1
Proposed Flow Stress Model Parameters for Lattices

Lattice Design	Density (kg/m³)	Model Parameters								
		A	m	B	n	C_1	C_2	C_3	C_4	p
Diamond	437.522	161.817	0.342	50.244	6.718	−0.054	−0.070	2.265	−0.037	1.285
I-WP	804.27	274.016	0.432	302.643	4.404	−0.041	−0.054	0.150	−0.004	1.011
Primitive	551.30	139.845	0.326	151.547	6.594	−0.038	−0.055	1.688	−0.005	1.132

The resulting model parameters for the three lattice designs are presented in Table 7.1. These parameters can be linked to aspects of the response curves. A influences the magnitude of the stress values most directly and is primarily focused on the stress value in the late plateau region. With all other parameters remaining constant, an increase in the value of A provides an upward shift in the response. The value of B alters the densification strain location, with an increase in B leading to a lower densification strain value. The parameter m is mainly linked to the slope of the response curve prior to the densification point. However, indirectly, this leads to a shift in the stress values from the onset of loading to the densification strain. Here, a lower value of m results in a flatter slope and higher stress value. n is responsible for the slope in the densification region, and a larger value for n leads to a steeper slope. For the strain rate sensitivity terms, a negative value indicates that the parameter value will decrease with an increase in strain rate. Lastly, the p-value indicates the change in the response due to changes in the thermal environment. As with the Johnson–Cook model, a higher value in this parameter indicates that an increase in temperature results in a softer material.

7.3   RESULTS AND DISCUSSION

7.3.1   QUASI-STATIC COMPARISON

The lattice designs were tested across six quasi-static strain rates: 0.001, 0.002, 0.005, 0.010, 0.050, and 0.100 sec⁻¹. Figure 7.9 depicts the quasi-static stress–strain

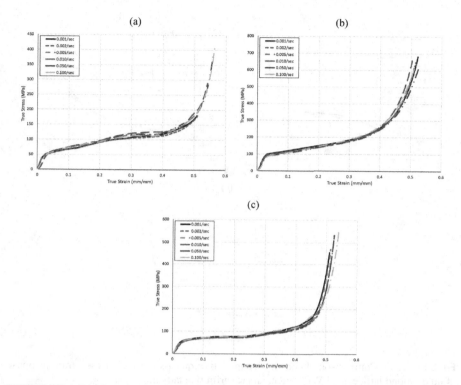

FIGURE 7.9 Quasi-static true stress–strain response curves of (a) diamond lattice, (b) I-WP lattice, and (c) primitive lattice.

responses for the three lattices, with the diamond lattice responses presented in Figure 7.9a, the I-WP lattice response curves in Figure 7.9b, and the primitive lattice responses in Figure 7.9c. These three figures show that there is little strain rate dependence within the quasi-static range. As such, for use in model comparison, the experimental data from the reference strain rate, 0.010 sec^{-1}, will be used for each lattice design.

Next, each of the described models was fit to the quasi-static compression testing data and compared quantitatively through the use of the mean square error between the experimental and model results, and qualitatively evaluating the closeness of the model fit to the data. Figure 7.10 shows the comparison of the model results to the experimental data for the three lattices.

The Rusch model, Avalle's updated model, and the new proposed model provide the most accurate models quantitatively for each lattice. Table 7.2 provides the mean square error results for each of the models under quasi-static compression. The Rusch model provides the closest fit to the experimental data for all three lattice designs. The proposed model is not far behind for both the diamond and primitive lattice responses, and it is still the third closest fit for the I-WP response. Avalle's 2019 model is also relatively close in fit to the Rusch and proposed model, and all three would be suitable for modeling the response. As a qualitative measure, all the

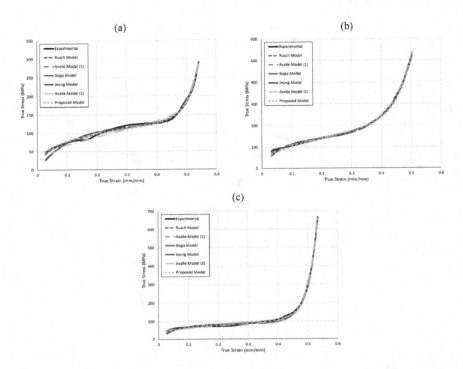

FIGURE 7.10 Comparison of constitutive models to quasi-static true stress–strain response of (a) diamond lattice, (b) I-WP lattice, and (c) primitive lattice.

TABLE 7.2

Mean Square Error of Constitutive Models under Quasi-Static Compression Loading

| Lattice | Mean Square Error | | | | | |
Response	Rusch	Avalle (1)	Goga	Jeong	Avalle (2)	Proposed
Diamond	18.32	41.76	45.12	42.15	45.48	25.06
I-WP	14.28	36.78	43.81	37.24	20.93	23.91
Primitive	45.89	79.17	86.47	77.37	56.53	46.88

models provided a reasonable approximation of quasi-static compression loading. Each model was developed with a mathematical formulation that is capable of representing the three response regimes through either a two- or three-term power law series. The strain rate–insensitive models have the advantage of being fit to a specific response curve, where the material was subjected to a single applied strain rate. This allows the model parameters to be more easily derived, as they do not have to be modified by a strain rate sensitivity term.

7.3.2 DYNAMICS COMPARISON

The lattice designs were tested across five dynamic strain rates: 200, 400, 600, 800, and 4,250 sec⁻¹. The first four tests were completed using the SHPB test, and the final test was conducted using the Taylor impact test. SHPB responses for the three lattices are shown in Figure 7.11. These response curves are notably different from the quasi-static response curves in magnitude, having a lower yield strength and plateau stress value, with a significantly shorter strain range. Nevertheless, there is a relatively consistent response across the tested strain rate range as with the quasi-static responses. Again, a single strain rate response, 800 sec⁻¹, will be used for comparison purposes.

As with the quasi-static responses, the strain rate–sensitive constitutive models were fit to the SHPB compression data and then compared qualitatively and quantitatively. Figure 7.12 shows the comparison between the model results and 800 sec⁻¹ experimental data for the three lattices.

Within the SHPB strain rate range, the new proposed model provided the closest representation to the experimental data for each lattice by both quantitative and qualitative measures. Avalle's strain-sensitive model also provided a good fit for each of the lattices, especially considering the I-WP response, and the Johnson–Cook model presented a reasonable response as well. These two models have the ability to

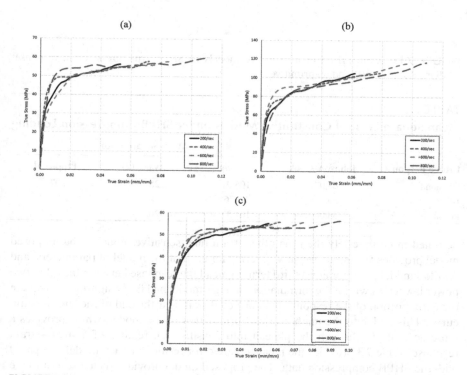

FIGURE 7.11 True stress–strain response curves from the SHPB for (a) diamond lattice, (b) I-WP lattice, and (c) primitive lattice.

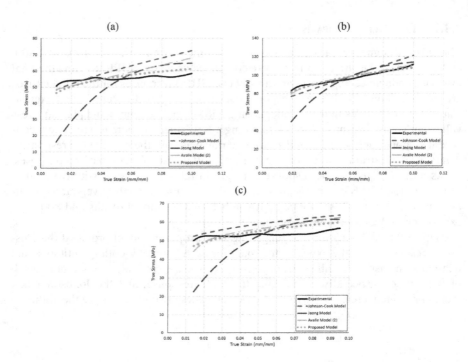

FIGURE 7.12 Comparison of constitutive models to SHPB true stress–strain response: (a) diamond, (b) I-WP, and (c) primitive.

TABLE 7.3

Mean Square Error of Constitutive Models under SHPB Compression Loading

		Mean Square Error		
Lattice Response	**Johnson–Cook**	**Jeong**	**Avalle (2)**	**Proposed**
Diamond	68.97	105.01	28.15	8.07
I-WP	39.01	83.76	4.17	3.09
Primitive	38.00	101.37	20.45	9.70

be tuned more precisely than the other strain rate–sensitive models. The proposed model provides strain rate sensitivity terms for each of the model fit parameters, and Avalle's model incorporates a third term to blend the response between the other two power law terms within the formulation. Jeong's model had fairly significant errors in the early portion of the response but corrected well near the end of the stress–strain curve. This is due to the mathematical formulation of the model, which provides a more accurate fit through the plateau into densification regions of a stress–strain response. Table 7.3 provides the mean square error results for each model compared with the SHPB compression data. The proposed model provides the closest fit to the experimental data for all three of the lattice designs. Both Avalle's updated model and the Johnson–Cook model provide a suitable fit for modeling the response as well. Because of the significant error early in the response for Jeong's model, it may not be suitable for modeling purposes depending on the strain range under consideration.

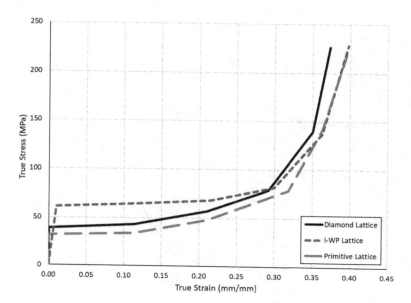

FIGURE 7.13 True stress–strain response curves under impact compression from the Taylor impact test.

The dynamic stress–strain responses for the three lattices under compressive impact loading from the Taylor test are depicted in Figure 7.13. Again, these response curves are markedly different from the previous quasi-static and SHPB response curves. These responses indicate a lower yield strength and plateau stress value, but here the strain range exceeded the densification strain, with each curve presenting plateau and densification regions. Due to the impact velocity and frame rate used by the digital camera, there were insufficient data to determine the linear elastic region of the stress–strain response accurately.

The strain rate–sensitive models were again compared with the experimental results from the Taylor impact test, Figure 7.14. At the 4,250 sec⁻¹ strain rate, the proposed model afforded the most accurate representation of the experimental data. As with the SHPB data, Avalle's strain-sensitive model provided a moderately good fit for each of the lattices, performing best when modeling the diamond lattice. Again, this is due to the number of tunable parameters in the models, which provides a better strain range and strain rate range for fitting. Jeong's model displayed the same early response errors, which led to a high overall error; however, in the case of the diamond lattice, the model fits the densification response almost exactly. As mentioned, each response reached the densification region at these strain rates, which led to the Johnson–Cook model being almost unusable since it is mathematically incapable of modeling the secondary stress rise.

Table 7.4 provides the mean square error results for each model compared with the Taylor impact test data. Once again, the proposed model provides the closest fit to the experimental data for each of the three lattice designs. Avalle's model still provides a suitable response fit for some lattice compression modeling, especially in the late plateau and densification regions. Due to the significant errors in the Jeong and Johnson–Cook models, they are likely unsuitable for modeling purposes.

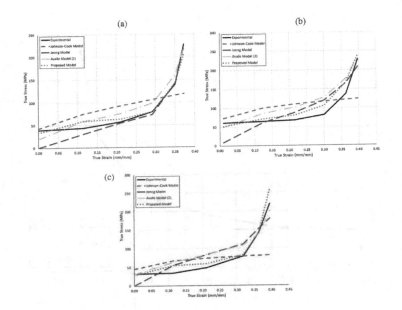

FIGURE 7.14 Comparison of constitutive models to Taylor impact true stress–strain response: (a) diamond, (b) I-WP, and (c) primitive.

TABLE 7.4

Mean Square Error of Constitutive Models under Impact Compression Loading

		Mean Square Error		
Lattice Response	Johnson–Cook	Jeong	Avalle (2)	Proposed
Diamond	2,432.89	260.38	253.81	110.89
I-WP	2,299.93	838.48	797.91	286.71
Primitive	3,542.17	770.28	476.88	203.68

REFERENCES

44. G. R. Johnson and W. H. Cook. A constitutive model and data for metals subjected to large strains, high strain rates, and high temperatures. In *Proceedings of the Seventh International Symposium on Ballistics*, pages 541–547, The Hague, 1983.
81. K. C. Rusch. Load–compression behavior of flexible foams. *Journal of Applied Polymer Science*, 13(11): 2297–2311, 1969.
82. K. C. Rusch. Energy-absorbing characteristics of foamed polymers. *Journal of Applied Polymer Science*, 14(6): 1433–1447, 1970.
84. M. Avalle, G. Belingardi, and A. Ibba. Mechanical models of cellular solids: Parameters identification from experimental tests. *International Journal of Impact Engineering*, 34(1): 3–27, 2007.
85. M. Avalle and G. Belingardi. A mechanical model of cellular solids for energy absorption. *Advanced Engineering Materials*, 21(4): 1–7, 2019.
125. V. Goga and B. Hucko, Phenomenological material model of foam solids, Journal of Mechanical Engineering, 65(1): 5–20, 2015.
126. K. Y. Jeong, S. S. Cheon, and M. B. Munshi. A constitutive model for polyurethane foam with strain rate sensitivity. *Journal of Mechanical Science and Technology*, 26(7): 2033–2038, 2012.

8 Lattice Damage Model

The Johnson–Cook (JC) damage model was modified for use with the lattice structures to model element damage and provide a basis for element deletion within the impact simulation. The basic model outline, as presented in Section 3.3.1, was followed with minor modifications. The following sections will outline the methodology used to obtain the relevant model parameters, with a discussion on the variations from the JC method.

8.1 DAMAGE MODEL PARAMETER DETERMINATION

The first step in determining the damage model parameters was incorporating the lattice constitutive model developed, as described in Chapter 7, into Abaqus. This required the use of an elastic-plastic model, using the Foam Hardening model that is part of the standard Abaqus package. A uniaxial compression model was developed within Abaqus that consisted of an upper and lower platen represented by rigid elements. The lower surface fixed and upper surface followed the same displacement control schedule used in experimentation and a specimen represented with deformable three-dimensional stress brick elements, C3D8R elements. The platen and specimen surfaces were tied together as master-slave contact surfaces during the simulation. General contact properties were also applied to the non-tied surfaces. Figure 8.1 depicts the pre-loading condition of the model.

Since the model was not able to be directly used in Abaqus, there was some variation in the stress–strain response under uniaxial compression, but it was not enough error to cause concern. Figure 8.2 shows the comparison between the experimental results and the response from the elastic-plastic model used in Abaqus. For each lattice, there is some disagreement between the curves in the late plateau region approaching the densification strain. However, there is good agreement of the model with the experimental results in the early plateau and densification regions.

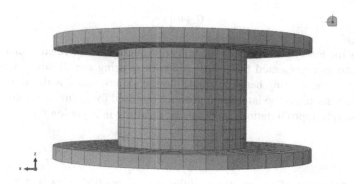

FIGURE 8.1 Uniaxial compression model developed in Abaqus.

DOI: 10.1201/9781003299639-8

139

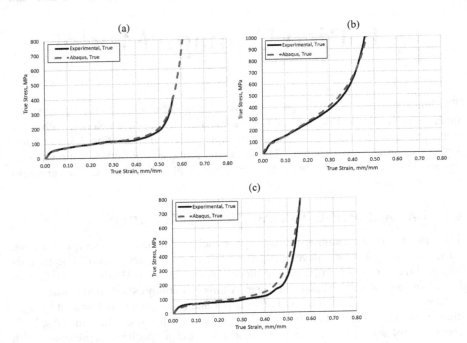

FIGURE 8.2 Comparison of Abaqus model to experimental results for the (a) diamond lattice, (b) I-WP lattice, and (c) primitive lattice.

The next step in the process was to calculate the failure stress so that the failure strain could be determined in the subsequent steps. As the lattice does not act as a continuum material, the failure stress was found using the surface energy concept [127]. The surface energy concept, used in conjunction with an atomic model, can provide an estimate of theoretical strength. The atomic model characterizes failure as the separation of the atoms within the material. With solid materials, the stress–strain response is approximated by a sine curve, Equation 8.1.

$$\sigma = \sigma_c \sin\left(\frac{2\pi x}{\lambda}\right) \tag{8.1}$$

where σ is the loading stress, σ_c is the failure strength, x is the relative displacement between atoms represented by $x = \alpha - \alpha_0$, with α being the distance between the atoms and α_0 the spacing between stable atomic planes, and λ is the wavelength of the approximation. As the inter-atom distance required for failure is relatively small, the small-angle approximation can be made, resulting in Equation 8.2.

$$\sigma = \sigma_c \frac{2\pi x}{\lambda} \tag{8.2}$$

Next, the definition of the elastic modulus can be applied, Equation 8.3, and the expression rearranged to provide Equation 8.4.

$$E = \frac{\sigma}{\varepsilon} = \frac{\sigma}{x/\alpha_0} \tag{8.3}$$

$$\sigma_c = \frac{\lambda E}{2\pi\alpha_0} \tag{8.4}$$

Surface energy, γ, is now defined as the amount of work performed to create a new surface through the breaking of atomic bonds. Using the sine approximation for the stress–strain response curve, the surface energy is equal to one-half of the area beneath the curve, represented in Equation 8.5.

$$2\gamma = \int_0^{\lambda/2} \sigma_c \sin \frac{2\pi x}{\lambda} = \frac{\lambda \sigma_c}{\pi} \tag{8.5}$$

Rearranging Equation 8.4 to solve for λ, then substituting the result into Equation 8.5 and solving for the failure stress, leads to Equation 8.6.

$$\sigma_c^2 = \frac{\gamma E}{\alpha_0} \text{ or } \sigma_c = \sqrt{\frac{\gamma E}{\alpha_0}} \tag{8.6}$$

γ is on the order of $0.01\ E\alpha_0$, which leads to the final form for the theoretical failure strength, Equation 8.7.

$$\sigma_c = \frac{E}{10} \tag{8.7}$$

This methodology remains valid for lattice materials, except failure of the lattice occurs in the densification region. Therefore, instead of using the modulus of elasticity as the numerator for Equation 8.7, the modulus for the densification slope is used, which provides Equation 8.8. The failure stress for each lattice design is presented in Table 8.1.

$$\sigma_c = \frac{E_D}{10} \tag{8.8}$$

Now that a failure stress has been determined and a compression model validated in Abaqus, the first three damage parameters can be calculated. The methodology

TABLE 8.1

Damage Model Parameters for Triply Periodic Minimal Surface –Based Lattice

Lattice Design	D_1	D_2	D_3	D_4	D_5	σ_f (MPa)
Diamond	0.278	0.520	0.475	−0.022	0.062	751.94
I-WP	0.169	0.407	0.176	−0.008	0.198	739.22
Primitive	0.214	0.451	0.207	−0.020	0.074	778.37

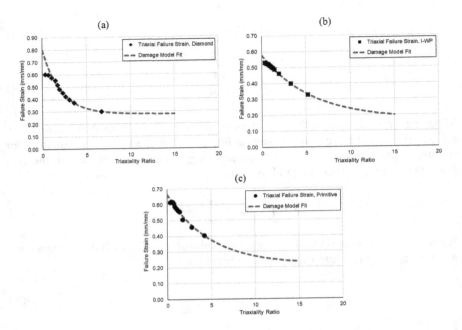

FIGURE 8.3 Determination of damage model parameters D_1, D_2, and D_3 for the (a) diamond lattice, (b) I-WP lattice, and (c) primitive lattice.

was carried out in Abaqus under the assumption that the lattice would behave in an isotropic manner [20]. The specimens were subjected to a uniaxial compression loading that matched the experimental tests with a bounding pressure added to create a triaxial loading condition. The bounding pressure was changed to achieve various triaxial stress states. From the stress response output from Abaqus, the failure strain was determined as the point at which the specimen reached the failure stress. The failure strain, ε_f, was then plotted against the triaxiality ratio, Q, and fitted with the expression presented in Equation 8.9. Figure 8.3 displays the charts for the three lattices.

$$\varepsilon_f = D_1 + D_2 e^{-D_3 * Q} \qquad (8.9)$$

The subsequent damage variable, D_4, is used to account for strain rate effects present within the lattice. Within Abaqus, the strain rate–sensitive model developed in Chapter 7 was used under uniaxial compression conditions from 0.001 to 3,000 sec^{-1}. The failure strain was found in a manner similar to that of the previous step, then plotted against the natural log of the strain rate ratio, Figure 8.4, with the slope of the linear regression providing the value for D_4.

The final damage parameter, D_5, is the temperature-sensitive damage term. The same constitutive model was used to find D_5, but a standard uniaxial compression load was applied under varying temperatures. The failure strain was found for each of the thermal conditions and plotted against the homologous temperature, Figure 8.5.

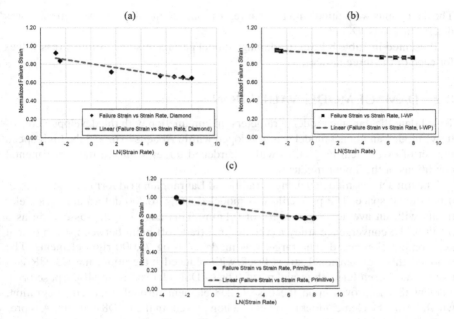

FIGURE 8.4 Determination of damage model parameter D_4 for the (a) diamond lattice, (b) I-WP lattice, and (c) primitive lattice.

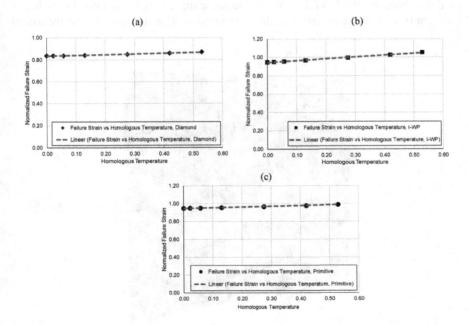

FIGURE 8.5 Determination of damage model parameter D_5 for the (a) diamond lattice, (b) I-WP lattice, and (c) primitive lattice.

The data points were fitted with a linear regression, and the slope of the regression was determined to be D_5.

Presented in Table 8.1 are the damage model parameters and failure stress values for each of the three lattices.

8.2   DAMAGE MODEL VALIDATION

To validate the damage model parameters, in conjunction with the developed constitutive model, an impact model was developed within Abaqus to model the high-speed impact of cylindrical projectiles with a hardened target to match the experimental conditions of the Taylor impact test.

The impact was modeled using a traditional Lagrangian grid formulation in three-dimensional space. The projectile was modeled using 10,700 deformable brick elements with an average edge length of 0.1 mm. This edge length was chosen as a result of the convergence study as the optimal trade-off point between strain output and computation speed. The target was modeled using 75,000 rigid elements. The 8-noded three-dimensional stress brick with reduced integration, the C3D8R element, was chosen for both model parts. The C3D8R element is an all-purpose brick element that improves upon the base C3D8 element through reduced integration, which limits the element locking phenomenon present in the C3D8 element. A representation of the Abaqus model is presented in Figure 8.6.

As a comparison, Figure 8.7 presents a side-by-side contrast of three points within the impact event for the Abaqus model, shown on top, and experimentation, shown on the bottom, for the I-WP lattice. Within these images, there is a good correlation between time increments and cylinder compression. The specimens were measured

FIGURE 8.6   Taylor impact test model developed in Abaqus.

(a)   (b)

(c)

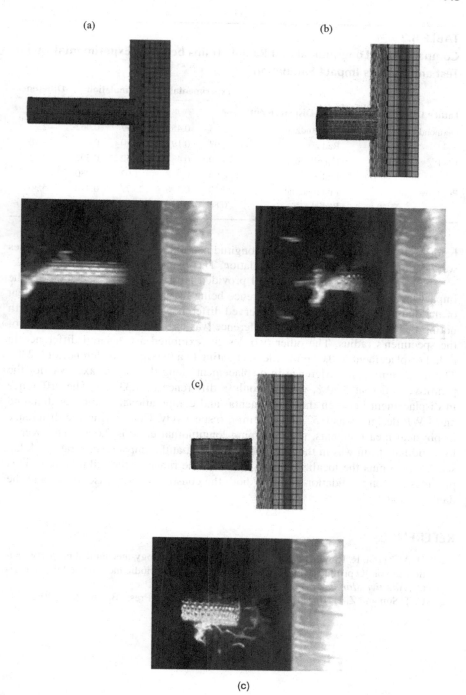

(c)

FIGURE 8.7 Comparison of Abaqus impact model to Taylor impact test results at various points: (a) pre-impact, (b) peak compression, and (c) post-impact.

TABLE 8.2

Comparison of Longitudinal and Radial Strains between Experimental Taylor Test and Abaqus Impact Simulation

		Experimental		Simulation		Differences
Lattice Design	Direction of Displacement	Δd	ε	Δd	ε	(%)
Diamond	Longitudinal	22.95	0.452	23.47	0.462	−2.27
	Radial	2.07	0.193	1.90	0.150	−8.18
I-WP	Longitudinal	24.70	0.487	25.52	0.503	−3.32
	Radial	0.99	0.078	1.02	0.080	3.03
Primitive	Longitudinal	25.10	0.496	23.77	0.470	5.30
	Radial	0.88	0.070	0.86	0.067	−2.27

before and after impact to determine longitudinal and radial strain, and these values were compared with those of the simulation. The results are presented in Table 8.2.

These results show that the model provided an accurate representation for the impact test, with the maximum difference being 8.18% for the radial displacement of the diamond lattice. While the observed difference equated to 8.18%, it is worth noting that the measured physical difference was only 0.17 mm, which is 1.32% of the specimen's radius. The other two designs exhibited a measured difference in radial displacement of 0.03 mm or less, equating to a maximum difference of 0.24%. The most significant difference in displacement along the impact axis was for the primitive lattice at 5.30%, which is only a difference of 1.33 mm. The difference in displacement between the experimental and computational results for diamond and I-WP designs was 0.52 and 0.82 mm, respectively. Considering the difference in physical measurements, the maximum longitudinal error is 2.62%. The overall longitudinal strain was in the range of 50% strain, at the impact strain rate of 4,250 sec^{-1}, which puts the localized stress values in the range of the failure stress. This provides a strong validation point for both the constitutive response model and the damage model.

REFERENCES

20. O. Al-Ketan, R. Rowshan, and R. K. Abu Al-Rub. Topology-mechanical property relationship of 3D printed strut, skeletal, and sheet based periodic metallic cellular materials. *Additive Manufacturing*, 19(January): 167–183, 2018.
127. C. T. Sun and Z.-H. Jin. *Fracture Mechanics*. Academic Press, Boston, MA, 2012.

9 Computational Modeling Techniques and Results

As a precursor to the final impact model, a three-dimensional, physics-based computational model was developed in Abaqus. This model was used to predict the damage and failure of both the projectile and target under high strain rate impact using the Johnson–Cook (JC) plasticity and damage models native in the Abaqus software. The materials and dimensions for the initial model were chosen to match experimental validation conditions [113, 128], to provide validation of the modeling configuration and simulation execution.

9.1 COMPUTATIONAL IMPACT MODEL

The initial computational model used for this study was a single projectile–single target assembly constructed in Abaqus using the inherent explicit finite element solver. The models all started at the point of impact with calculated impact velocities from the experimental data. Contact between the two parts was modeled using the general contact algorithm native to Abaqus, which uses a penalty method to impose contact constraints through introduction of increased local stiffness. The general contact algorithm was used to enforce contact between two bodies, and model friction between parts. This algorithm allows for automatic contact definition based on surface inclusion. Within the Lagrangian systems, contact forces are generated based on node, face, and edge interactions. It is also capable of enforcing contact between Eulerian and Lagrangian systems, compensating for any discrepancies between the two constructs. Of the contact algorithms available within Abaqus, general contact is the only contact algorithm that can be used with three-dimensional models and is capable of evaluating across a mixed-model-type simulation. The friction developed here follows the Coulomb friction model, which formulates the friction coefficient based primarily on contact pressure for impact but includes surface slip and temperature at the contact point [43].

The projectile was configured with a cylindrical body of 24.7 mm in length and 16.7 mm in diameter to develop an equivalent system to the experimental setup. It incorporated a blunt nose geometry and was modeled using reference material properties of 6061-T6 Aluminum, shown in Table 9.1, along with the JC parameters shown in Table 9.2. The impact velocity of 970 m/s was applied as a load to the rear face of the projectile, with no other boundary conditions enforced on the projectile within the simulation.

The targets were configured as square-faced plates that were 203 mm by 203 mm with a thickness of 12.7 mm, using the same material set. The target had fixed

DOI: 10.1201/9781003299639-9

TABLE 9.1

Material Properties of 6061-T6 Aluminum

Material Property	Value
Elastic modulus, E	69 GPa
Poisson's ratio, ν	0.33
Density, ρ	2,700 g/m^3

TABLE 9.2

Johnson–Cook Model Parameters for 6061-T6 Aluminum

Johnson–Cook Parameter	Value
Yield stress, A	324.1 MPa
Strain hardening parameter, B	113.8 MPa
Strain rate parameter, C	0.002
Thermal softening exponent, m	1.34
Strain hardening exponent, n	0.42
Damage constant, D_1	−0.77
Damage constant, D_2	1.45
Damage constant, D_3	−0.47
Damage constant, D_4	0.0
Damage constant, D_5	1.6

boundary conditions applied at both the upper and lower surfaces, as if it were affixed in a mount, with the other edges left as free surfaces. A depiction of the simulation boundary and initial condition is shown in Figure 9.1.

All the simulations were evaluated for an impact time of 12 μs. Presented here are the validation cases, using a variety of finite element analysis (FEA) techniques, compared with the previously acquired experimental results as means for evaluation of the modeling techniques [128]. Three variations of the base model were developed to evaluate different modeling techniques for use in this work. The first two models were based solely on Lagrangian FEA techniques, comprising a full-scale model to match the real-world dimensionality of the experiment and a symmetry-reduced model that used dual-axis symmetry to achieve a quarter-scale model. The third incorporated the use of smoothed particle hydrodynamics (SPH) to model the target. The target was chosen for the case of SPH as it would see larger deformations than the projectile.

Figure 9.2a presents the stress colormap that corresponds to traditional Lagrangian models of Figures 9.3–9.5, and Figure 9.2b presents the stress colormap scales that correspond to the mixed models of Figures 9.6 and 9.7.

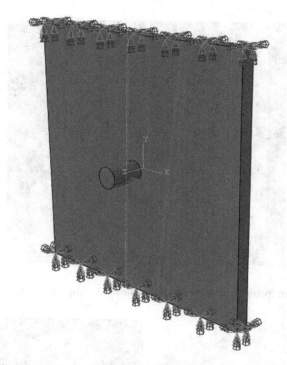

FIGURE 9.1 Depiction of impact simulation loading and boundary conditions.

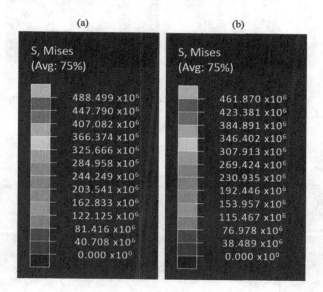

FIGURE 9.2 Stress colormap scale: (a) traditional Lagrangian model and (b) mixed traditional–smoothed particle hydrodynamics model.

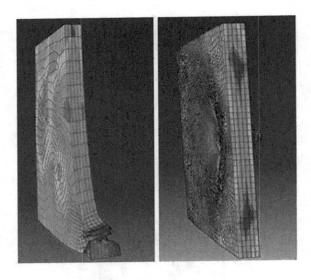

FIGURE 9.3   Symmetry-reduced versus full-scale model impact model.

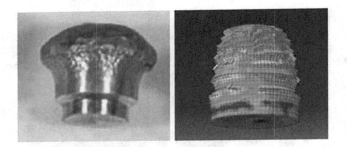

FIGURE 9.4   Visual comparison of the projectile with the finite element results. (With permission from L.E. Schwer.)

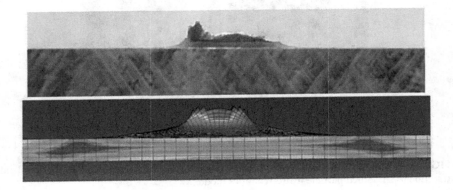

FIGURE 9.5   Visual comparison of the target with the finite element results. (With permission from L.E. Schwer.)

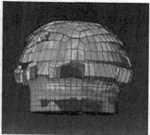

FIGURE 9.6   Visual comparison of the target with the smoothed particle hydrodynamics result. (With permission from L.E. Schwer.)

FIGURE 9.7   Visual comparison of the projectile with the smoothed particle hydrodynamics results. (With permission from L.E. Schwer.)

9.2   FULL LAGRANGIAN MODELS

The projectile and target were both symmetric across two axes, allowing the model to be cut along those axes and reduced in scale, to a quarter scale of the full model. This was only possible because of the shared planes of symmetry about the impact location of both the parts, forces, and boundary conditions. A visual comparison of the symmetry-reduced and full-scale model is provided in Figure 9.3.

As mentioned, in both the reduced and full-scale models, the C3D8R eight-noded stress bricks with reduced integration were used as the elements for analysis of the projectile and target. In the full-scale model, the projectile incorporated 7,800 elements, and the plate target was comprised of 50,000 elements. The number of elements chosen for the parts was based on a convergence study performed by varying the element size from 1/50th down to 1/500th of the plate width using the symmetry-reduced model. To save on computational run time, the largest element size that still provided consistent results was chosen, which was 1/100th of the plate width. The JC plasticity and failure models were used to estimate the element failure modes of the materials, with element deletion occurring for cells with equivalent plastic strain values greater than 1.0. This value was found to most closely match the model data to previous experimental results. Figures 9.4 and 9.5 show a comparison between the full-scale model deformation results against the experimental results of the projectile and target plate, respectively [128].

There is a noticeable difference in the projectile results shown in Figure 9.4. The most significant aspect of this difference, showing the leading face of the projectile narrowing versus mushrooming, is due to inaccurate equivalent plastic strains for the impact face elements combined with the element deletion scheme used. At the projectile impact velocity, and subsequent strain rate, the traditional Lagrangian modeling technique is subject to mesh distortion causing errors, which are compounded through the CDTI methodology. Ultimately, this will lead to inaccuracies in the element strain and equivalent plastic strain, which would cause errant element deletion.

The symmetry-reduced model provided nearly identical results as the full-scale model yet took approximately one-third of the time to run, 0.3 processor hours versus 1.0 processor hours for the full model.

9.3   MIXED SMOOTHED PARTICLE HYDRODYNAMICS– LAGRANGIAN MODEL

While the deformation results of the traditional Lagrangian model appeared to match the experimental results adequately for the target, the deformation and residual velocity of the projectile were not well modeled. Residual velocity is the projectile's velocity on exit of the target. Therefore, a mixed SPH–Lagrangian model was developed to evaluate the projectile and target dynamics and interactions more closely. As previously mentioned, it was decided that the target would be modeled using SPH techniques, as it would be subject to larger deformations than the projectile based on the deformation seen in the traditional Lagrangian grid model and experimentation. The projectile was still modeled using the traditional C3D8R element with the same seeding as the full Lagrangian model, using 7,800 elements. The plate was discretized into SPH particles to match the size of the elements used in the full Lagrangian model, resulting again in 50,000 elements used to model the target. However, due to the use of ghost particles in modeling the boundary interaction within the SPH methodology, 100,000 particles were ultimately used in the computational analysis, which comprised 50,000 particles used to represent the target plate and 50,000 particles used to model the boundary interaction throughout the impact scenario. Figure 9.6 depicts the SPH target deformation following impact under the same parameters as the traditional model compared with the experimental results.

In this figure, the particles that are no longer attached to the target plate would have been removed from the target during impact and shown as deleted under the traditional FEA method. As shown here, the SPH model more closely replicates the asymmetric shearing around the exit hole that was found in the experimental results. While the impact problem is described as a symmetrical problem, there are potential sources for the asymmetry in the simulation, such as asymmetric discretization of a part and numerical round-off. The projectile was discretized the same between the two models, and some element asymmetry was noted in both the impact and rear faces. The asymmetry in the projectile discretization can still provide physically relevant data, as it can be seen as a similar effect to imperfections within the part, or nonhomogeneity within the material or structure. As different mathematical methodologies are used in the two models, an asymmetric discretization could lead to asymmetry in the mixed model but not in the traditional model. Similarly, numerical round-offs could cause asymmetry in either model, but as different equations are used, the round-off would likely manifest differently between the models.

While in this model, only the target was converted into an SPH model, the use of SPH in the assembly also gave a better appreciation of the projectile response throughout the interaction. Figure 9.7 shows the projectile following impact compared with the experimental projectile deformation.

Using this model, the projectile shows a deformation pattern more similar to the experimental results than with the previous modeling technique. A primary contributor to the accuracy of the solution is the more precise displacement solution of the target through the use of the SPH technique. With contact prescribed throughout much of the simulation, the displacement solution of the target has a direct impact on the forces imposed on the projectile, which in turn will dictate the plastic strain accumulation used in the damage model. An additional element that leads to the closeness in results is due to the equivalent plastic strain value used for element deletion being tuned more specifically for the projectile in the SPH–Lagrangian model than for the traditional model while maintaining a nearly identical target failure pattern. However, the SPH model took significantly longer to process than either the full-scale or symmetry-reduced models, with a run time of 5.2 processor hours.

9.4 FURTHER COMPARISON OF MODELS

As mentioned above, the traditional grid model did not provide an adequate result for the projectile's residual velocity, but the SPH model was able to very closely match the results seen through experimentation; see Table 9.3. Also shown here are the computer processing times required for each model.

TABLE 9.3
Comparison of Computational Models

Model	Residual Velocity (m/s)	Residual Velocity Error (%)	Processor Time (hour)
Experimental	336.194	-	-
Traditional, full	30.186	91.021	1.0
Traditional, reduced	30.189	91.020	0.3
Smoothed particle hydrodynamics	335.406	0.234	5.2

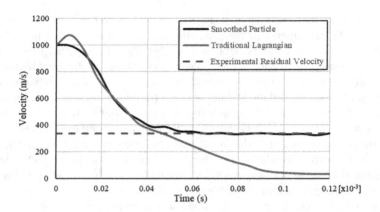

FIGURE 9.8 Comparison of projectile residual velocity between traditional Lagrangian finite element and smoothed particle hydrodynamics models.

Figure 9.8 shows the velocity plot of the projectile for both the traditional model and SPH model against the simulation time, with $t=0$ being initial contact. The velocity values are taken from elements along the center line of the projectile. As shown here, the SPH velocity follows a smooth and expected deceleration from the initial impact velocity to the projectile's residual velocity. On the other hand, the traditional grid velocity shows an initial acceleration within the first time step, then decelerates more quickly down to a velocity roughly one-tenth of that observed in experimentation.

The initial increase in velocity is likely due to the high strain rate of impact, which could not be accurately modeled by the traditional explicit methodologies. One of the most essential differences between the two methods is that SPH is meshless, and the problem domain is discretized with particles that do not have a fixed connectivity. Thus, large displacement problems are better evaluated since there is no need to evaluate the internal forces based on individual volume integration as required in the traditional approach. The traditional Lagrangian method requires a continuity of nodes, which requires the integration of the volume represented by the element geometry, and under large deformations, it may be so distorted that the evaluation will produce errors in balancing force distribution. At this impact velocity, and subsequent strain rate, the full Lagrangian model would have produced some error in the elemental volume integration required to determine the internal force vector that is used to determine the time step acceleration term. In the SPH method, there is no need to evaluate the integration of volume within the element as there are no element-connecting nodes, rather the internal force vector is through a pre-established association with the neighboring particles using the kernel function. This association is predetermined and becomes part of the derivative included within the conservation equations. Another potential source of error that led to the lower residual velocity is the implementation of the Coulomb friction coefficient in Abaqus's general contact algorithm. If the friction coefficient is too high, it can lead to binding in the model as it progresses. While the same friction coefficient was used in both models, the differences in the relative motion of the projectile and target between the models would change the application of the friction and lead to errors in the velocity.

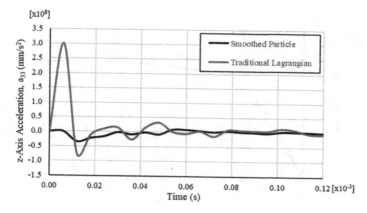

FIGURE 9.9 Comparison of impact-axis acceleration between traditional Lagrangian finite element and smoothed particle hydrodynamics models.

Figure 9.9 presents the impact axis, or z-axis, acceleration for a target particle and node on the edge of the initial contact. For reference, the model was oriented with initial impact velocity along the negative z-axis. The acceleration of the traditional Lagrangian element depicts a significantly larger rise in positive acceleration than the SPH particle in the beginning of the response. This difference is an important factor in the initial increase in velocity shown by the full Lagrangian model and why the initial velocity of the SPH model stayed constant. The acceleration response of the full Lagrangian model also exhibits larger peaks and troughs, which is indicative of the errors manifesting in the internal energy volume integration.

Furthermore, a comparison between the internal energy versus time for the two models is presented in Figure 9.10. Overall, the trend between the two curves is similar to the traditional model displaying higher internal energy values, see Figure 9.10a, but there is a unique artifact within the traditional model early within the simulation run; see Figure 9.10b. This variation within the internal energy curve is likely due to an error developed within the internal force calculations of the traditional method as mentioned above. Since the traditional Lagrangian model is based on Newton's second law, the error in the acceleration derived from the internal force calculation would have been carried forward to the velocity and displacement vectors through the CDTI methodology shown in Equations 2.1–2.3 and compounded throughout the time step integration process. This is likely the cause of the significant error in residual velocity realized by the two traditional models.

While both modeling techniques rely on a Lagrangian reference frame according to the internal interactions and external forces and thus evolve the system in time, within SPH, the mathematical process of satisfying the three conservation equations alongside the equation of state reduces the likelihood of error. To highlight the differences in computed displacements between the two methods, a comparison of element strain over time is presented in Figure 9.11. Figure 9.11a compares the element strain of an element on the impact face of the projectile. As seen in the figure, both curves follow the same trend, although the SPH model strain is roughly twice that of

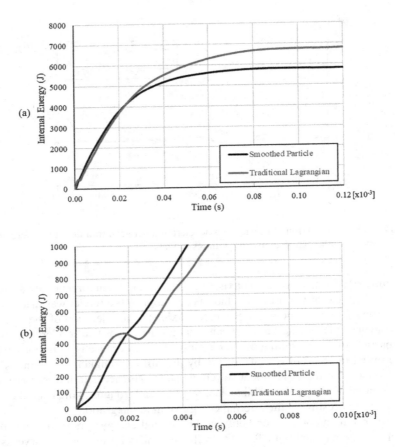

FIGURE 9.10 Model internal energy versus time comparison: (a) first run duration and (b) initial reaction.

the traditional model. Figure 9.11b shows a similar comparison for an element on the rear face of the projectile through the simulation duration. For this case, the response curves are not quite in alignment, although the general trend of the strain over time is comparable. The back face element shows the opposite case of the front face, in that here the strain values of the traditional model are higher than that of the SPH model, by roughly 70%. These figures emphasize the resultant difference in nodal displacements, and ultimately projectile strain, between the two models used.

Figure 9.12 depicts the plastic strain accumulation of a single target element or particle on the outer edge of the initial contact between the particle and target. The JC Damage model, which is used to determine element and particle failure, relies on the plastic strain accumulation within the model. The SPH plastic strain accumulation was characterized by a smooth accumulation up to a maximum strain of 0.73 mm/mm. The traditional Lagrangian response was significantly more chaotic, displaying several discontinuities. The element reached a maximum plastic strain of 1.1 mm/mm, although the mean plastic strain in the plateau region was 0.85 mm/mm. The more

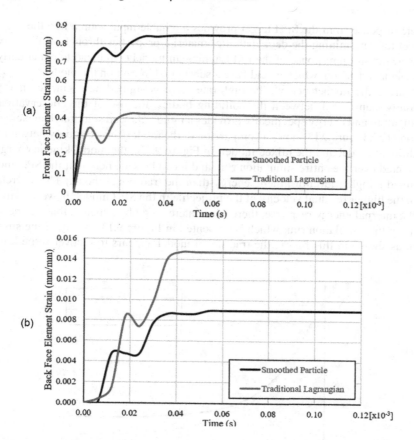

FIGURE 9.11   Element strain versus time: (a) projectile impact face and (b) projectile back face.

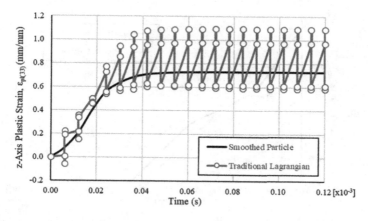

FIGURE 9.12   Comparison of plastic strain along the impact axis between traditional Lagrangian finite element and smoothed particle hydrodynamics models.

stable response from the SPH target provides a reliable input source into the damage model for determining the degradation of material properties throughout the impact. The erratic strain response of the full Lagrangian model could lead to either early or late element deletion, which would have a substantial impact on the model.

During the impact period, the high-rate force is applied, and kinetic energy is partially transferred between the colliding bodies, the use of the conservation of momentum and energy specifically provides a more balanced solution than Newton's second law. Figure 9.13 shows a comparison of the model kinetic energy between the traditional model and the mixed model. In Figure 9.13a, the model kinetic energy is presented over the entire simulation run, and it can be seen here that the SPH model retained a higher level of kinetic energy than the traditional model. This correlates with the higher residual velocity of the projectile in the SPH model; however, similar to the internal energy response, there is an interesting phenomenon that can be seen early in the simulation run, which is presented in Figure 9.13b. Early in the simulation, as shown in this figure, the traditional model appears to recover some kinetic

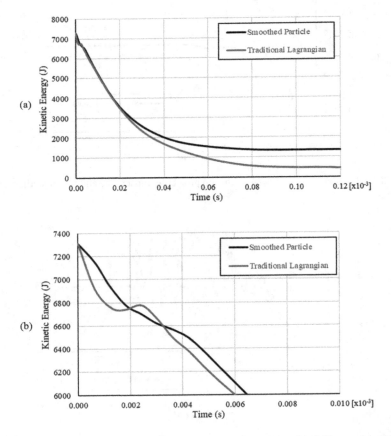

FIGURE 9.13 Model kinetic energy versus time comparison: (a) full run duration and (b) initial reaction.

energy, whereas the mixed model does not show this anomaly. This peculiar feature is a further indication of the errors produced in the Lagrangian impact model.

REFERENCES

43. D. Systèmes. Abaqus version 6.12 theory manual, 2016.
128. L. E. Schwer. Aluminum plate perforation: A comparative case study using LaGrange with erosion, multi-material ale, and smooth particle hydrodynamics. *7th European LS-DYNA Conference*, Salzburg, Austria, pages 1–27, 2009.
113. D. G. Spear, A. N. Palazotto, and R. A. Kemnitz. Modeling and simulation techniques used in high strain rate projectile impact. *Mathematics*, 9(3) 1–29, 2021.

10 Projectile Impact Results

This section presents the results of the impact testing of projectiles that incorporated a lattice core. Experimental testing was accomplished on additive manufacturing projectiles as described in Section 3.3.2. Additionally, computational simulations were run for the same models and a solid Inconel 718 (IN718) projectile.

10.1 EXPERIMENTAL PROJECTILE IMPACT

Three different projectile models were fabricated from IN718 for experimentation. They were precision-turned following fabrication to have exact dimensions for use within the pneumatic cannon. Each of the models was 12.7 mm in diameter and 63.5 mm in length and included a different lattice design within the core section. The three lattices used were the diamond, I-WP, and primitive triply periodic minimal surface–based designs. The relative density of the lattice sections was set to maintain a constant projectile mass, so that the kinetic energy and momentum between projectile models would be the same during testing. The chosen relative density was 25%. The actual relative densities and mass for each projectile can be found in Table 10.1. Each of the projectiles indicates a higher actual relative density than designed, which is likely due to trapped powder within the projectile shell-to-lattice interface.

Based on the previous testing accomplished with the pneumatic cannon, a chamber pressure of 1,200 psi was chosen to achieve an impact velocity of 250 m/s for the mass of the projectiles. The impact velocity was determined through digital image correlation, using the camera frame rate, and reference distance provided a length per pixel within the image. The actual reservoir chamber pressure and impact velocity for each of the test events are listed in Table 10.1. The resultant velocities were slightly lower than expected, which is in all likelihood due to the differences in mass and lower chamber pressures.

TABLE 10.1

Experimental Conditions of Projectile with Lattice Core

Lattice Core	Lattice Relative Density (%)	Projectile Mass (g)	Chamber Pressure (psi)	Impact Velocity (m/s)
Diamond	25.28	50.06	1,198	247.20
I-WP	25.07	49.66	1,185	549.44
Primitive	25.58	50.65	1,181	244.53

DOI: 10.1201/9781003299639-10

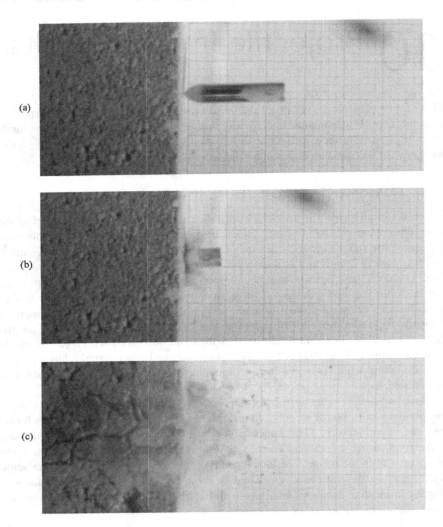

FIGURE 10.1 Ogive projectile impact sequence: (a) initial contact of the projectile and the target, (b) half-penetration of the projectile, and (c) full-penetration of the projectile.

A depiction of the impact event is shown in Figure 10.1. The projectile in this image sequence is the projectile that included the diamond lattice core. Within this figure, the projectile can be seen penetrating the target with minimal damage occurring before the projectile has reached half-length penetration of the target. In Figure 10.1(c), more pronounced debris back spray from the penetration can be observed, along with significant cracking present on the observed face of the concrete target.

Similar impact conditions were seen for each of the experimental runs, with the projectile impacting mid-target and perpendicular to the target face. The penetration depth for each of the experimental runs is presented in Table 10.6. The diamond core projectile displayed the furthest penetration, with a depth of 87.68 mm, with the primitive core projectile having a lower penetration distance at 78.09 mm. The penetration depth of the I-WP core projectile was markedly less than the other two runs at 67.32 mm. This difference in results was due to the projectile hitting a large piece of aggregate within the concrete target. These results indicate that the diamond core provides the best condition for impact based on penetration depth.

The dimensional changes for the projectiles were also taken as measures of performance. Three metrics were chosen for analysis: the change in length, the change in mid-length diameter, and the change in bulge diameter. The results are shown in Table 10.7. For the change in projectile length, the diamond projectile provided a significantly lower value than the other designs. None of the specimens exhibited a significant change in mid-length diameter, and the difference in the change in bulge diameter was relatively insignificant. This outcome is a potential indication that the diamond lattice core absorbed more energy during the impact through internal deformation, as it penetrated deeper into the target with minor overall deformation.

10.2 COMPUTATIONAL PROJECTILE IMPACT

In the fundamental calculations and assessment of the penetration testing event, it was concluded that there would be significant deformation of the target at the impact velocity planned. Based on these preliminary estimates of damage and the initial impact model results, described in Chapter 9, a mixed smoothed particle hydrodynamics (SPH)-Lagrangian formulation was used to model the impact between the lattice-inclusive projectile and concrete target. As the target model was projected to achieve greater displacements within the analysis, the decision was made to represent this part with particles. At the same time, the projectile still used traditional grid methods. The overall model design is described in Section 3.3.2, and the conditions of the finite element analysis simulation were designed to match those of the experimental testing. The model setup is presented in Figure 10.2. As with the initial model, the general contact algorithm was used due to the selection of a mixed model.

Because of the geometry of the projectile shell and lattice core sections, a hex-dominated mesh scheme was used under the advancing front algorithm in Abaqus. This methodology led to the primary use of three-dimensional stress brick elements

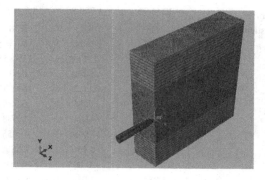

FIGURE 10.2 Projectile-concrete impact model setup in Abaqus.

of a linear geometric order, the C3D8R element, from the Abaqus Element Library. The C3D8R element is a general-purpose element with a single integration point. The reduced integration attribute allows the element to surmount the shear locking issue common in the full-integration element, C3D8, when used under high-plasticity scenarios. Shear locking is a known problem of all first-order, full-integration solid elements because the element is unable to exhibit pure bending modes [43]. Additionally, three-dimensional stress tetrahedral elements of a quadratic order were used, the C3D10M element. The C3D10M element is a modified 10-noded quadratic tetrahedron element that performs well for deformation analysis and minimizes the potential for shear and volumetric locking. The chosen meshing strategy led to 6,289 nodes and 5,355 elements being used to model the projectile, including 4,755 C3D8R elements and 600 C3D10M elements. The concrete target was discretized into SPH particles to match element geometries with the projectile. The target's core used a refined mesh to provide better detail of the stress wave and deformation present within the immediate impact region. Based on the element sizes used, there were 113,400 particles used to represent the target. The number of elements and particles chosen for the parts was based on a convergence study performed by varying the average element size of the projectile from 1/60th down to 1/1,000th of the projectile length. The largest element size that still provided consistent results was chosen, which was 1/100th of the projectile length. Along with the full-scale model, a symmetry-reduced model was developed to reduce computation time and provide an improved visual representation of the projectile's interior.

The elastic response of the lattice section was modeled using a linear relationship with the calculated modulus of elasticity, and the newly developed constitutive model for lattices was used to model the plastic response. The IN718 portions of the projectiles were represented with a linear elastic response and the Johnson–Cook (JC) plasticity model. The material properties of the lattices are presented in Table 10.2.

TABLE 10.2
Material Properties Used in Computational Impact Model

Material Property	Diamond	I-WP	Primitive	IN718
Density, $\rho\left(\text{kg/m}^3\right)$	1778.92	1,779.82	1,777.91	7,970.07
Modulus of elasticity E (GPa)	2.25	3.35	2.28	49.56
Poisson's ratio, ν	0.07	0.04	0.03	0.29
A (MPa)	161.82	177.98	139.85	965.30
M	0.34	0.43	0.33	1.18
B (MPa)	50.24	211.23	151.55	1,370.00
N	6.72	4.40	6.59	0.65
C_1	−0.054	−0.041	−0.038	0.017
C_2	−0.070	−0.054	−0.055	-
C_3	2.265	0.150	1.688	-
C_4	−0.037	−0.004	−0.005	-
P	1.285	1.011	1.132	-

TABLE 10.3

Damage Model Parameters Used in Computational Impact Model

Damage Parameter	Material			
	Diamond	I-WP	Primitive	IN718
D_1	0.278	0.169	0.214	0.110
D_2	0.520	0.407	0.451	0.750
D_3	0.475	0.176	0.207	−1.450
D_4	−0.022	−0.008	−0.020	0.040
D_5	0.062	0.198	0.074	0.890

Both the lattice and solid metals used the JC damage model to account for element degradation and failure. The damage model parameters for the lattice sections and IN718 are depicted in Table 10.3.

The response of the concrete target was represented with a Holmquist–Johnson–Cook (HJC) model, and the damaged crack plasticity was used to account for particle degradation, with the property values shown in Table 3.1.

The boundary conditions of the target were enforced to emulate the conditions of the experimental test, with fixed boundaries represented on the side faces, back face, and bottom surface. These boundary conditions led to the conditions of no allowable displacement or rotation along the outer edges of the target. The upper surface of the target had a uniform load applied to model a restraining load placed on the target that provided near-fixed conditions. Therefore, unlike the other surfaces, the upper surface was allowed some movement, and the elements were permitted to rotate. The projectile had an initial velocity applied to the entire part but was otherwise unconstrained. A depiction of the boundary conditions applied to the assembly of parts for the symmetry-reduced model is shown in Figure 10.3. The symmetry-reduced model incorporated additional boundary conditions to mirror the assembly across the cut face. As the restraining load presented near-fixed boundary conditions, the symmetry-reduced model and full model results were nearly identical.

The simulation time required for the projectiles to reach zero forward velocity within the model was 0.004 sec, and a maximum time increment of 1×10^{-7} sec was used.

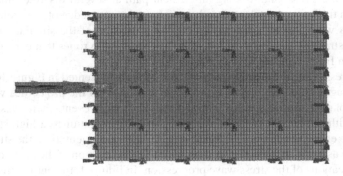

FIGURE 10.3 Applied boundary conditions to projectile-concrete impact model in Abaqus.

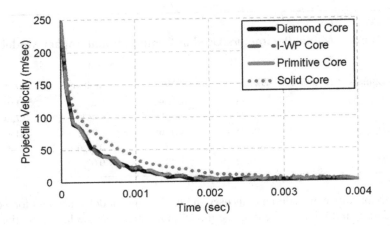

FIGURE 10.4   Projectile velocity over time during impact with concrete target.

The velocity profiles for the four simulations are presented in Figure 10.4. Due to the increased mass of the solid projectile, the impact velocity was decreased to match kinetic energy at impact. The three lattice core projectiles followed a similar deceleration profile; however, the solid projectile initially displayed a more gradual deceleration during target penetration. The dissimilarity in deceleration is due to the difference in momentum, which is given by mass times velocity. The solid projectile having a higher mass had a higher momentum than the lattice core projectiles. The impulse provides the change in momentum, or mass times change in velocity; therefore, a greater force over time is required to slow and stop the solid projectile.

A progression of the impact event is reproduced in Figure 10.5. This progression starts at the beginning of the simulation, showing the initial time step of the contact between the projectile and target, then the half-length penetration of the projectile, and the full-length projectile penetration depth. These images align close to the same time increments as in Figure 10.5.

One of the unique features of finite element simulations is the ability to evaluate the stress distribution along the length of the projectile. This was accomplished by acquiring the stress values along an element path at several discrete time steps, or frames. In this case, the selected path was along the outer edge of the projectile. Five time steps were chosen, focusing on the initial portion of the simulation as that is when the stress field is the most dynamic. The simulation times that correspond with the chosen frames are displayed in Table 10.4.

The stress along the element path at these times is presented in Figure 10.6 Shown in this image is the diamond core projectile, although all the projectiles with lattice cores displayed similar results. Throughout the impact event, there was continual contact with the ogive portion of the projectile, this maintained a high stress value during the entire simulation. Depending on the element orientation, the stress values are shown as being in tension or compression. The aft portion of the element path displays indications of the stress wave progression. In Frame 1, the entire stress field aft of the mid-point is in compression; however, looking at the stress field for Frame 2,

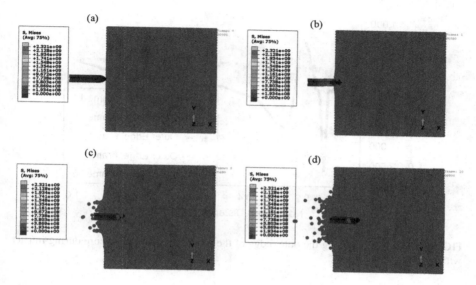

FIGURE 10.5 Projectile impact simulation sequence: (a) start of simulation (Frame 0), (b) initial contact of the projectile and the target (Frame 1), (c) half-length penetration of the projectile (Frame 5), and (d) approximate full-length penetration of the projectile (Frame 10).

TABLE 10.4

Corresponding Simulation Times for Frame Number

Frame Number	Simulation Time (sec)
1	8.0×10^{-5}
2	1.6×10^{-4}
3	3.2×10^{-4}
9	7.2×10^{-4}
30	2.4×10^{-3}

between 60% and 90% projectile length, the stress field is in tension. The section that is in tension is indicative of the reflection wave returning off the back face of the projectile. The later frames show that the projectile returns to a state of slight compression, and then returns to a neutral stress state by Frame 30, which is just over halfway through the simulation.

Figure 10.7 shows the same stress field incrementation as in Figure 10.6, only for the solid core projectile. As with the lattice core projectiles, the stress field at Frame 1 is primarily in compression outside of the ogive nose section. However, for the solid core projectile, the stress field displays lower overall stress values, especially aft of 20% projectile length. There are a few locations that indicate that elements are in tension further back in the projectile, which again indicates the progression

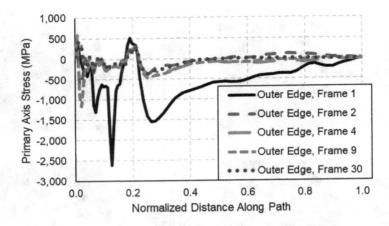

FIGURE 10.6   Stress along the outer edge of the projectile with lattice core during impact with concrete target.

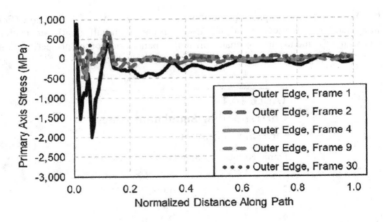

FIGURE 10.7   Stress along the outer edge of the projectile with solid core during impact with concrete.

of a stress wave. Here the stress wave appears to endure later in the simulation than for the lattice core projectiles. The stress field for the Frame 9-time increment of Figure 10.7 still shows a small, but defined, wave between 80% and 100% length, yet Figure 10.6 shows a flutter response that remains in compression starting with Frame 4. This observation suggests that the lattice core aids in the dissipation of the energy associated with the stress wave.

In addition to evaluating the outer edge of the projectile, the stress field of the projectile's centerline within the core section was also assessed. The same frames and corresponding time increments used in evaluating the outer edge were used here. Figure 10.8 shows the stress fields of the four projectile designs for comparison at these times. As with the projectile shell, the nose section of the core maintains a

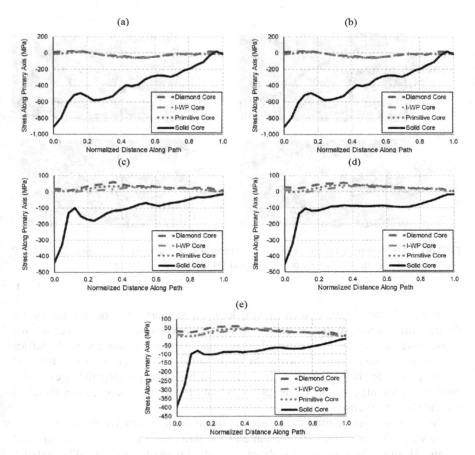

FIGURE 10.8 Stress along the centerline of the projectile core: (a) Frame 1 (8.0×10^{-5}), (b) Frame 2 (1.6×10^{-4}), (c) Frame 4 (3.2×10^{-4}), (d) Frame 9 (7.2×10^{-4}), and (e) Frame 30 (2.4×10^{-3}).

higher stress load than the remaining portion of the core, and as time progresses, the stress maintained in the aft portion of the core trends toward a nearly constant value. The stress field of the solid core remains in compression throughout the entire simulation, except for Frames 1 and 2, where there is a small portion of the core in tension near 95% core length. The tensile region shown in these two images is presumably due to a stress wave reflection from the back face of the core, where it meets the centerline channel.

All lattice cores display similar trends among themselves but are in sharp contrast to the solid core. Frames 1 and 2 for the lattice cores show a stress field that is primarily in compression or very close to a neutral stress state. However, starting in Frame 9, the stress field of the lattice cores are almost solely in tension. In all likelihood, this is due to the radial expansion occurring near the bulge region, which is at approximately one-third of the core length. The deformation that is taking place in

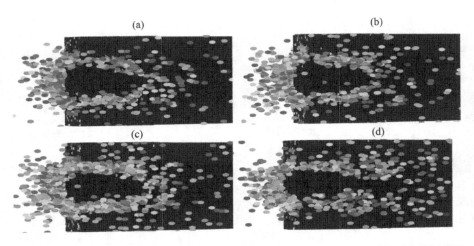

FIGURE 10.9   Damage imparted on concrete target by (a) diamond core projectile, (b) I-WP core projectile, (c) primitive core projectile, and (d) solid core projectile.

the core, indicated by the tension stress field, is the mechanism of energy absorption for the lattice structure, just as cellular deformation leads to the plateau region in a quasi-static compression response. However, under these circumstances, the lattice core section is no longer supporting the impact load, which now increases the load on the projectile shell. This leads to a trade-off between reducing the stress wave throughout the aft portions of the projectile and increasing the stress on the shell around the core section.

In addition to evaluating the projectile, the amount of damage imparted on the target was also considered. The volume of the damaged material was determined within Abaqus. The target was isolated from the projectile, and the SPH particles were overlaid on a target grid representation to determine the material loss region. Figure 10.9 shows the overlay for each of the four projectile impacts.

The intersection of the particles and grid was used to outline the bounding damage volume, as shown in Figure 10.10. Three lattice core projectiles do not exhibit the same penetration depth as the solid core projectile; however, the targets of the lattice core projectiles indicate that more damage occurred on the impact face of the target than with the solid core projectile. Of the three lattice core designs, the diamond core projectile presented the most damage on the impact face, with 70.09% more impact face damage than the solid core projectile.

The corresponding target damage volumes and percent of target damaged are presented in Table 10.5. The solid core projectile caused the greatest amount of damage to the target, with a total damage volume of $3.84 \times 10^{-4}\,\text{m}^3$. Of the three lattice core projectile designs, the diamond core projectile exhibited the most significant amount of damage being imparted to the target, $3.32 \times 10^{-4}\,\text{m}^3$ in total damage volume. The solid core projectile is 15% heavier than the other three projectile designs, so as a means of comparison between the projectiles and to account for the variation in mass, a specific damage parameter was determined. The specific

(a)

(b)

(c)

(d)

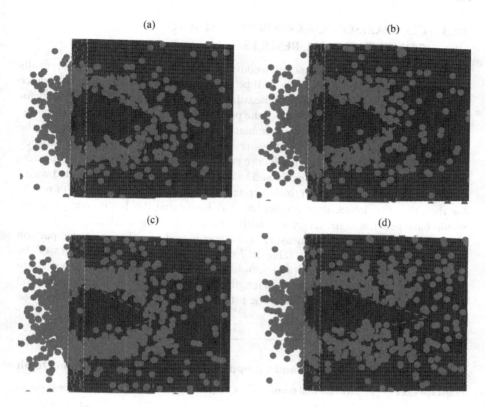

FIGURE 10.10 Outline of concrete target damage volume caused by (a) diamond core projectile, (b) I-WP core projectile, (c) primitive core projectile, and (d) solid core projectile.

TABLE 10.5

Damage Imparted to Concrete Targets by Projectile Core Design

Lattice Core	Target Damage Volume (m³)	Target Damage (%)	Projectile Mass (kg)	Specific Damage (m³/kg)
Diamond	3.32×10^{-4}	7.93	3.04×10^{-2}	1.09×10^{-2}
I-WP	2.82×10^{-4}	7.19	3.04×10^{-2}	9.92×10^{-3}
Primitive	2.28×10^{-4}	5.44	3.04×10^{-2}	7.50×10^{-3}
Solid	3.84×10^{-4}	9.16	3.48×10^{-2}	1.10×10^{-2}

damage was calculated as the damage volume divided by the projectile mass. Even taking into account the projectile's mass, the solid core projectile caused the greatest amount of damage, but the diamond core projectile produced 99% of the specific damage that the solid core projectile caused. This finding indicates that a trade-off is possible within the specific damage of a projectile between penetration depth and impact face damage.

10.3 COMPARISON OF EXPERIMENTAL AND COMPUTATIONAL RESULTS

To validate the computational model, a comparison between the experimental results and simulation must be made. This will be done using the measure of performance outlined in Section 3.3.2. The first measure of performance that can be used for comparison is the penetration depth of the projectile. Table 10.6 displays the penetration depths for the experiments and simulations. The results were quite close for the diamond and primitive core projectiles, remaining within a 5% difference between the model and live testing. The I-WP core projectile showed a significant difference, 23.31%; however, this is easily explained by the fact that the projectile collided with a large piece of aggregate within the concrete target. Based on the closeness of results for the other two lattice core projectiles, it is likely that the I-WP core projectile would have penetrated the target to a depth near the simulation value.

The second performance measure to be considered as a basis for comparison is the projectile's deformation. Table 10.7 displays the dimensional changes of the projectiles under experimental and simulation conditions. The three deformation changes of interest were the change in length, change in mid-projectile diameter, and change in bulge diameter. For all of the projectiles, the change in length was close

TABLE 10.6

Comparison of Experimental and Computational Projectile Penetration Depth

Projectile Core	Experimental (mm)	Computational (mm)	Difference (mm/%)
Diamond	87.68	83.40	−4.28/−4.88
I-WP	67.32	83.01	15.69/23.31
Primitive	78.09	79.01	0.92/1.18
Solid	-	118.3	-

TABLE 10.7

Comparison of Experimental and Computational Projectile Dimensional Change

Measure of Performance		Diamond	I-WP	Primitive	Solid
				Projectile Core	
Change in length	Experimental (mm)	0.18	0.30	0.39	-
	Computational (mm)	0.20	0.35	0.36	0.10
	Difference (mm/%)	0.02/13.70	0.05/18.12	−0.03/−9.10	-
Change in diameter	Experimental (mm)	0.00	0.00	0.01	-
	Computational (mm)	0.001	0.001	0.002	0.000
	Difference (mm/%)	0.001/100	0.001/100	−0.008/−600	-
Change in bulge	Experimental (mm)	0.04	0.03	0.04	-
	Computational (mm)	0.10	0.07	0.09	0.04
	Difference (mm/%)	0.06/60.98	0.04/58.66	0.05/55.91	-

between the experimental and simulation results. While the percentage differences range from 9.10% to 18.12%, the actual measured differences are minor. The projectile with the I-WP core exhibited the largest difference between experimentation and simulation with a difference of 0.05 mm. Considering the length of the projectile, this represents a 0.08% difference in longitudinal length. None of the projectiles exhibited a meaningful change in the mid-projectile diameter, with only the primitive core projectile having a measurable change in diameter in experimentation. As with the change in length, the change in bulge diameter has larger percentage difference values, ranging from 55.91% to 60.98%, but the measured difference is small. The maximum difference in bulge diameter between experimental and computational results was 0.06 mm, for the diamond core, which equates to only a 0.05% change in the radius of the projectile. Furthermore, in each case, the experimental impact presented a lower bulge than simulations predicted.

REFERENCE

43. D. Systèmes. Abaqus version 6.12 theory manual, 2016.

11 Conclusions

11.1 SUMMARY OF CONCLUSIONS

A review of past research into additively manufactured cellular structures, specifically triply periodic minimal surface (TPMS)–based lattice structures, revealed several shortcomings and the need for further investigation into characterizing these materials for use in computational modeling. This led to a joint experimental and analytical approach to this research. The initial quasi-static testing of lattice specimens showed that the guidance provided by the American Society for Testing and Materials (ASTM) E9-19 was not adequate for characterizing additively manufactured periodic structures using the medium-length test specimen parameters. Modified test specimen geometries and aspect ratios were used in further experimentation, showing that a one-to-one length-to-diameter ratio is optimal for determining the compressive response through the three regions of a lattice stress–strain curve [109]. Based on the initial results of column and cube specimens, additional testing was accomplished by looking at the deformation behavior and mechanical response of three TPMS-based lattices, along with several design variations. Imagery taken during quasi-static compression testing showed deformation patterns of different lattices. All the designs indicated a mixed mode of deformation between stretching-dominated and bending-dominated deformation. The diamond and primitive designs displayed strong indications of single or double shear failure, whereas the I-WP designs had a uniform failure along horizontal rows of cells. A clear relationship between relative density and mechanical performance was further highlighted within testing. The derivative designs developed through mathematical manipulation did not perform as well as the base designs [107]. The results from the derivative design led to an evaluation of topology optimization being applied to the three TPMS lattice designs. Across all three optimized cell designs, the optimized cell saw an improved stress field response when considering the reduction in cell volume and subsequent reduction in mass. Even without considering the mass reduction, the diamond and I-WP lattice cells showed improved performance. The I-WP lattice saw the most improvement through the topology optimization process, with a 62.63% decrease in maximum stress loading along with a 29.98% reduction in cell volume. Overall, topology optimization proved beneficial for designing TPMS-based lattice cells under an initial uniaxial loading condition. While a constraint on the optimization process was that the cells retain the ability to be replicated into sheet lattices for use in engineering applications, there was some question as to how the optimized cells would perform when replicated, so the base lattice designs were carried forward. Though the optimization study was not presented, it was found that a deeper investigation into the influence of lattice topology, cell size, cell density, and surface thickness was performed using statistical analysis techniques. The results revealed statistically significant effects attributable to all four factors when considering the mechanical response of the specimens under uniaxial compression. Least squares estimate of the factor effects indicated that the lattice surface thickness provides the most significant impact, and the cell density of the specimens provides the most negligible impact on the subsequent material properties. Furthermore, the observation of multiple factor combinations achieving the same material property results opens up the trade space between these factors within the design stage. This finding will allow for primary and secondary effects to be considered when designing a lattice structure, especially in energy-absorbing applications [114]. Split Hopkinson Pressure Bar testing was conducted to determine the dynamic characteristics of the lattice structures. There was a significant difference in the mechanical response of the lattices between the quasi-static and dynamic response regimes; however, within the quasi-static and tested dynamic ranges separately, the mechanical response indicated an insensitivity to the applied strain rate. The trends observed remained consistent between the two regimes, showing plastic deformation behavior for the diamond and primitive lattices and exhibiting a strain hardening behavior for the I-WP lattice.

Another prominent outcome of this testing was that, unlike the response of the base material, the lattice designs showed a decrease in elastic modulus, yield strength or flow stress, plateau stress, and toughness as the strain rate increased. The Johnson-Cook (JC) Flow Stress model served as a reasonably practical model of the plastic stress and strain response of the three chosen lattice structures over the quasi-static and dynamic response regimes through the plateau region to approximately 30% strain. Having a usable plasticity model will allow these lattice designs to be included in the computational analysis of both quasi-static and dynamic scenarios up to the initial onset of densification.

To model the lattice response beyond densification, significant modifications would have to be made to the JC Flow Stress formulation, ultimately leading to the development of a new plasticity model. To account for the densification portion of the lattice response, a novel flow stress model, which includes strain rate and temperature sensitivity terms, was developed to describe the mechanical response of metal lattice designs. This model was refined out of previous mechanical models, namely the Rusch and JC models. All the proposed model parameters were able to be derived from experimental compression testing data, with a repeatable process being developed and applied to each lattice design. The new model was applied under three different strain rate bands and was able to properly model the compression behavior of three different lattice designs manufactured out of Inconel 718 (IN718). The model was also compared with previously developed constitutive models for cellular materials, comparing favorably against the other models when considering the dynamic material response. The goodness of fit from the proposed model is primarily due to the tunability of the model, especially considering the four strain rate sensitivity terms. The result is a model that has proven capable of accurately representing the mechanical response of cellular lattices across a strain rate range of 0.001–$4{,}250$ sec^{-1} using three different testing schemes. Due to the high strain rates and large deformations present in high-velocity impacts, it was imperative that a damage model be determined for use in modeling.

A variation on the JC fracture model was developed using a surface energy approximation for the failure stress of the lattices and an inverse methodology based on the work of Banerjee [86]. This technique relied on the use of an elastic-plastic model within Abaqus that could be validated against experimental results. The JC damage model parameters could then be determined by changing the simulation and material model conditions. The resultant damage model was then compared with the results of the Taylor impact tests for each of the three lattice designs under consideration with favorable results.

Two computational modeling techniques, a traditional Lagrangian grid and mixed Lagrangian-smoothed particle hydrodynamics, were considered in modeling high-velocity impact to determine optimal high strain rate finite element analysis simulation methodologies. These two techniques show that a trade-off must be made in the modeling of high-velocity projectile impact between computational cost and simulation performance. Considering the complexities of three-dimensional modeling, the Lagrangian model does an adequate job modeling physical deformation even at nearly 1,000 m/s and at a reasonable computational cost. However, there are drawbacks in Lagrangian analysis. The most notable deficiency is that it does not model larger deformations or displacements well, which was evident in the residual velocity results being one-tenth of that seen in experimentation. The mixed model was capable of handling higher strain rates and larger deformations better, which was evidenced in the closeness of the simulation results to the experimental values. The mixed model was able to achieve a residual velocity within a quarter of a percent of the experimental results and also displayed asymmetric results in the deformation pattern similar to experimentation. While the computational cost was five times that of the more traditional Lagrangian technique, the improved accuracy in the solution makes the mixed model preferable for simulating high strain rate projectile impact.

With the required material models developed and finite element analysis simulation methodologies determined, a computational model was created to evaluate the lattice, as a part of a later projectile, under impact conditions with a concrete target. In addition, projectiles incorporating the lattice core were fabricated and tested at an impact velocity of approximately 250 m/s. Two performance measures were used to validate the model: penetration depth and projectile deformation. Considering the closeness in results between the experimental cases and simulations, the computational model developed provides an adequate representation of the real-world conditions. This allowed the computational model to be used to analyze the stress field during the impact event. The analysis showed similar stress field results for each of the three lattice core designs, but the diamond core achieved the best overall results. The solid projectile still performed better based on the measures of performance. However, the lattice core appeared to decrease the propagation and continuation of the impact stress wave compared with the solid projectile. Furthermore, evaluating the total damage volume imparted on the concrete target, the diamond core projectile caused an equivalent amount of damage to the solid core projectile. These results further indicate that the elastic-plastic stress–strain relationship provided by the newly developed constitutive model and the damage model based on the JC failure model can be incorporated into Abaqus and provide valuable results for lattice structures and combined lattice-solid components under highly dynamic scenarios.

Based on the results from the computational model, the inclusion of a lattice core will aid in the dissipation of the stress wave generated by an impact, but the projectile is likely to undergo more deformation. Of the three lattice core designs considered, the diamond core yielded the best overall results considering the chosen performance metrics. The reduction in mass due to the inclusion of any lattice core means that the projectile will have less momentum during the impact resulting in a lesser penetration depth than a solid projectile, even when considering similar kinetic energy at impact. However, if the penetration depth is secondary to the amount of damage imparted by the projectile, the diamond core projectile is capable of causing the same amount of damage as a solid projectile with a reduced mass.

REFERENCES

86. M. Avalle and G. Belingardi. A mechanical model of cellular solids for energy absorption. *Advanced Engineering Materials*, 21(4): 1–7, 2019.

107. D. G. Spear, A. N. Palazotto, and R. A. Kemnitz. Mechanical properties of additively manufactured periodic cellular structures and design variations. *Journal of Engineering Materials and Technology*, 143(4), 05 2021. 041004.

109 D. G. Spear, A. N. Palazotto, and R. Kemnitz. Survivability and Damage Modeling of Advanced Materials. In *Survivability: Papers Presented at the AIAA SciTech Forum and Exposition 2020*, pages 123–134. American Institute of Aeronautics and Astronautics, 1 2020.

114. D. G. Spear and A. N. Palazotto. Investigation and statistical modeling of the mechanical properties of additively manufactured lattices. *Materials*, 14(14): 3962, 2021.

Appendix A
TPMS Lattice Structure Generation Code, MATLAB

```
%%%%%%%%%%%%%%%%%%%%%%%%%%%%%%%%%%
% TPMS Lattice Generator
%%%%%%%%%%%%%%%%%%%%%%%%%%%%%%%%%%

% This program generates a TPMS based lattice through a chosen
% design space
% based on user defined cell properties and design space
% parameters, then
% outputs the design as an .stl file for 3D manufacturing.

clear all; close all; clc;

%% Cell Properties
% User design choices for size of the individual cells and
% overall build
% space are input and then basic calculations are performed
% that will be
% used in lattice generation.

% User Design Inputs
cs = 3+1/3; % TPMS Cell Size in mm (Dimensional Length of a
% Unit Cell)
cd_w = 3; % Cell Density through Build Width (Number of Cells
% through the Build Width)
cd_h = 3; % Cell Density through Build Height (Number of Cells
% through the Build Height)
t = 0.25; % TPMS Surface Thickness in mm (Sheet Thickness of
% the Generated Lattice)
f = 20; % Cell Fineness Parameter (Number of Divisions within
% a Unit Cell)

% Calculations
hs = cs/2; % Half Cell Size in mm
w = cs*cd_w; % Overall Build Width in mm
hw = w/2; % Build Half-Width in mm
h = cs*cd_h; % Overall Build Height in mm
ht = t/2; % Half Thickness of Lattice Sheet
m = pi/hs; % Periodicity Multiplier
```

```
% Design Space Range
xl = -hw; xu = hw;
yl = -hw; yu = hw;
zl = -hw; zu = hw;

%  Design Space Range to Grid
x = [xl:cs/f:xu];
y = [yl:cs/f:yu];
z = [zl:cs/f:zu];

% Design Space Grid
[X,Y,Z] = meshgrid(x,y,z);

%% TPMS Surface Selection
% Select the desired lattice design from the TPMS options
% below by
% uncommenting the 3 corresponding lines below the TPMS design
% name.  The
% top line is the trigonometric representation of the TPMS
% cell, the second
% line is the function evaluation throughout the design space,
% the final
% line is the name that the .stl file will be saved as in the
% parent
% directory.

% Diamond Surface Function
% DMS = @(x,y,z) sin(m*x/2).*sin(m*y/2).*sin(m*z/2) + sin(m*x/
% 2).*cos(m*y/2).*cos(m*z/2) + cos(m*x/2).*sin(m*y/2).*cos
% (m*z/2) + cos(m*x/2).*cos(m*y/2).*sin(m*z/2);
% Trigonometric Representation
% F = DMS(X,Y,Z); % Function Evaluation
% stlname = ['D_build.stl']; % File name

% Gyroid Surface Function
% GMS = @(x,y,z) sin(m*x/2).*cos(m*y/2) +
% sin(m*y/2).*cos(m*z/2) + sin(m*z/2).*cos(m*x/2);
% Trigonometric Representation
% F = GMS(X,Y,Z); % Function Evaluation
% stlname = ['G_build.stl']; % File name

% I-WP Surface Function
% IMS = @(x,y,z)  2*(cos(m*x).*cos(m*y) + cos(m*y).*cos(m*z) +
% cos(m*z).*cos(m*x)) - cos(2*m*x).*cos(2*m*y).*cos(2*m*z);
% Trigonometric Representation
% F = IMS(X,Y,Z); % Function Evaluation
% stlname = ['I_build.stl']; % File name

% Lidinoid Surface Function
% LMS = @(x,y,z) 0.5*(sin(2*m*x).*cos(m*y).*sin(m*z) +
% sin(2*m*y).*cos(m*z).*sin(m*x) +
```

```
% sin(2*m*z).*cos(m*x).*sin(m*y)) - 0.5*(cos(2*m*x).*cos(2*m*y)
% + cos(2*m*y).*cos(2*m*z) + cos(2*m*z).*cos(2*m*x)) + 0.15;
% Trigonometric Representation
% F = LMS(X,Y,Z); % Function Evaluation
% stlname = ['L_build.stl']; % File name


% Primitive Surface Function
PMS = @(x,y,z) cos(m*x) + cos(m*y) + cos(m*z); % Trigonometric
% Representation
F = PMS(X,Y,Z); % Function Evaluation
stlname = ['P_3x3_10mm.stl']; % File name


%% Create Thickened Cell
% Applies surface offset to TPMS lattice surface to generate
% desired
% surface thickness.

% Generate Central Surface based on Chosen TPMS Design
Surf = isosurface(x,y,z,F,0);


% Create Offsets for Desired Thickness based on Surface
% Normals
SNorms = isonormals(x,y,z,F,Surf.vertices); % Surface Normals
% at Vertices
SNunit = SNorms./vecnorm(SNorms,2,2); % Surface Unit Normals
% at Vertices
OSoffset = Surf.vertices + ht*SNunit; % Outer Surface Offset
ISoffset = Surf.vertices - ht*SNunit; % Inner Surface Offset


% Find Free Boundary Surfaces in order to Close Lattice
% Surface
TRfb = triangulation([Surf.faces; Surf.faces + max(max(Surf.
faces))],[ISoffset; OSoffset]); % Inner
% &
% Outer Surface Triangulation
FB = freeBoundary(TRfb); % Surface Free Boundaries


% Enclose Openings Between Inner and Outer Surfaces
FBtr = FB; % Surface Free Boundary Index
hFB = length(FB)/2; % Half of the Free Boundary Index
FBtr(1:hFB,3) = FB(hFB+1:end,1); % Inner Surface to Outer
% Surface Triangulation
FBtr(hFB+1:end,3) = FBtr(1:hFB,2); % Outer Surface to Inner
% Surface Triangulation


% Combine Surfaces Arrays into Single Array for Triangulation
sVerts = [ISoffset; OSoffset]; % Inner and Outer Surface
% Vertices
sFaces = [Surf.faces; Surf.faces + max(max(Surf.faces))];
```

```
% Inner and Outer Surfaces Faces
sFull = [sFaces; FBtr]; % Add Boundaries

%% Output of Design
% The final arrays are triangulated into a mesh then saved to
% an .stl file
% for use in 3D manufacturing.

% Triangulation of Build for .stl Output
TR = triangulation(sFull,sVerts);

% Output Design to */stl
stlwrite(TR,stlname);

trisurf(TR,'FaceColor',[0.9 0.9 0.9]);
```

Index

Note: **Bold** page numbers refer to tables; *italic* page numbers refer to figures.

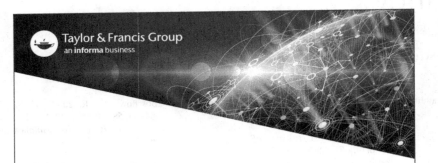

Printed in the United States
by Baker & Taylor Publisher Services

Printed in the United States
by Baker & Taylor Publisher Services